T0205571

Innovative Polymeric Adsorbents

Kyoichi Saito · Kunio Fujiwara
Takanobu Sugo

Innovative Polymeric Adsorbents

Radiation-Induced Graft Polymerization

 Springer

Kyoichi Saito
Chiba University
Chiba
Japan

Takanobu Sugo
KJK Co., Ltd.
Takasaki
Japan

Kunio Fujiwara
KJK Co., Ltd.
Takasaki
Japan

ISBN 978-981-13-4186-1   ISBN 978-981-10-8563-5   (eBook)
https://doi.org/10.1007/978-981-10-8563-5

Preface

The recovery of useful components or the removal of harmful components is essential for sustaining the world's resources and environments. Adsorption methods are economically feasible and environmentally friendly to capture useful and harmful ions and molecules using specified adsorbents. However, conventional adsorbents such as ion-exchange resins and activated carbon have limitations in terms of form and functionality. By radiation-induced graft polymerization, we have prepared various forms of polymeric adsorbents that include ion-exchange, chelate-forming, and affinity moieties and that immobilize extractants, enzymes, and inorganic compounds to satisfy the requirements for emerging applications and to realize high performance.

Some of our polymeric adsorbents and materials have been manufactured and used commercially. This book covers an extensive scope of topics on separation: the removal of radioactive ions at TEPCO's Fukushima Daiichi Nuclear Power Plant, the recovery of noble metal ions, the purification of bioproducts, enzyme immobilization, and the electrodialysis of seawater.

Researchers and engineers engaged in separation of ions or molecules always have a continuing interest in novel adsorbents. In this book, adsorbents with new forms and high performance are presented on the basis of scientific elucidation of graft-chain behavior.

Chiba, Japan   Kyoichi Saito
Takasaki, Japan   Kunio Fujiwara
Takasaki, Japan   Takanobu Sugo

Acknowledgements

Polymeric trunk polymers are essential starting materials in the preparation of novel polymeric adsorbents by radiation-induced graft polymerization. We appreciate the generosity of Asahi Kasei Chemicals Co., Mitsubishi Rayon Co., now Mitsubishi Chemical Co., INOAC Corporation, and TAMAPOLY Co. for supplying various trunk polymers such as porous hollow-fiber membranes, porous sheets, and non-porous films.

K. Saito worked in the Department of Chemical Engineering at the University of Tokyo from 1982 to 1994, and in 1994 joined the Department of Applied Chemistry and Biotechnology at Chiba University. All results and discussion described in this book were derived from the continuous efforts of many students of the two universities. The students were fascinated by the magic of radiation-induced graft polymerization, and they devised both preparation schemes for polymeric adsorbents and evaluation methods for their performance. In particular, we thank the following students previously and currently enrolled in doctoral courses at the two universities for their contribution to scientific findings and technological inventions: Takahiro Hori, Hideyuki Yamagishi, Kazuya Uezu, Min Kim, Satoshi Tsuneda, Satoshi Konishi, William Lee, Kei Kiyohara, Noboru Kubota, Hidetaka Kawakita, Kaori Saito, Shiho Asai, Katsuyuki Sato, Akio Iwanade, Kazuyoshi Miyoshi, Nobuyoshi Shoji, Ryo Ishihara, Kyohei Hagiwara, Yuichi Shimoda, Koji Miyauchi, Kunio Fujiwara, Takato Harayama, and Tsuyoshi Nagatani.

We thank Tsuyoshi Yoshida, Junichi Kanno, Satoshi Tsuneda, Takashi Yoshikawa, Kazuyoshi Miyoshi, Ryo Ishihara, Shiho Asai, Yuta Sekiya, Akio Iwanade, Satoshi Umino, Yuya Hirayama, Daiki Kudo, Masaki Iwazaki, and Shoko Naruke for preparing the fine graphics and illustrations. We also thank Dr. Yuji Hazeyama for his exhaustive review of our English manuscript. Our thanks go to Mrs. Michiko Hamamoto, our secretary, for editing the manuscript. Our appreciation also goes to Dr. Daisuke Umeno and Dr. Shigeko Noma-Kawai in our laboratory for valuable discussion. Finally, we would like to express our appreciation to Dr. Shinichi Koizumi and Ms. Asami Komada of Springer Japan for their patience and support.

Chiba, Japan   Kyoichi Saito
Takasaki, Japan   Kunio Fujiwara
Takasaki, Japan   Takanobu Sugo
December 2017

Contents

Chapter 1
Fundamentals of Radiation-Induced Graft Polymerization

Abstract Among the various graft polymerization methods, radiation-induced graft polymerization is powerful in that various forms of existing polymers can be selected as trunk polymers and converted into polymeric adsorbents. In particular, preirradiation graft polymerization has an advantage that the graft polymerization step can be separated from the irradiation step, which will enhance the industrial production of graft-type materials. The grafting of an epoxy-group-containing vinyl monomer, glycidyl methacrylate, enables the introduction of different functional moieties such as ion-exchange and chelate-forming groups, and hydrophobic and affinity ligands. In this chapter, batch and flow-through modes are described as methods of evaluating the performance of adsorbents for metal ions and proteins.

Keywords Preirradiation graft polymerization · Glycidyl methacrylate
Ion-exchange group · Chelate-forming group · Affinity ligand

1.1 Technical Terms

1.1.1 Graft Polymerization

An analogy between grafting performed by a gardener and that by a radiation chemist is illustrated in Fig. 1.1. Grafting is an effective agricultural technique for gardeners and farmers. When a tree that is resistant to severe climate and poor soil bears little or no fruit on its branches, the gardener forms a graft site by cutting a branch with scissors. Then, he/she grafts a branch capable of abundantly producing, for example, succulent fruit.

Apples in Aomori Prefecture of Japan are produced from trees propagated using the grafting technique. A branch yielding delicious apples is joined to a healthy trunk tree at grafting sites, and then, the tree will produce delicious apples. Similarly, chemists produce radicals by irradiating a trunk polymer with electron beams and gamma rays to form grafting sites. Then, a polymer branch with functional capabilities is grafted onto these sites of the trunk polymer. Graft

© Springer Nature Singapore Pte Ltd. 2018
K. Saito et al., *Innovative Polymeric Adsorbents*,
https://doi.org/10.1007/978-981-10-8563-5_1

1

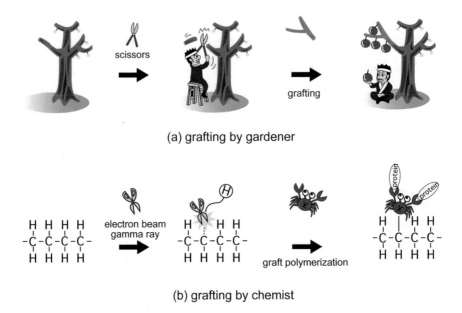

Fig. 1.1 Principle of grafting

polymerization enables us to prepare desirable and useful materials for separation and reaction.

Excitation sources for producing radicals by graft polymerization include radiation, plasma, light, and chemicals. Among graft polymerization procedures using these sources, radiation-induced graft polymerization is superior to other grafting techniques because a high density of electron beams or gamma rays can generate a large number of radicals in polymers of various components and forms [1, 2]. The components of polymer include polyethylene (low density, high density, and ultra-high molecular mass), polypropylene, polytetrafluoroethylene, and nylon 6 (Fig. 1.2a). The forms include porous hollow-fiber membranes, nonwoven fabrics, films, porous sheets, fibers, and particles (Fig. 1.2b).

Some chemical bonds must be destroyed by excitation to form radicals in the trunk polymer as the starting points of graft polymerization. The successive contacts of vinyl monomers, i.e., double-bond-containing monomers, with the radicals in the trunk polymer extend a polymer chain before termination occurs. The resultant polymer chain grafted onto the trunk polymer is referred to as a graft chain or a grafted polymer chain.

An appropriate combination of component and form will provide us with a new class of adsorbents. For example, an ion-exchange polymer chain grafted onto the surface of a porous hollow-fiber membrane provides high-rate binding of proteins because of negligible diffusional mass-transfer resistance of the proteins to the ion-exchange group, promoted by the permeative flow of a protein solution through

Fig. 1.2 Components and forms of trunk polymer

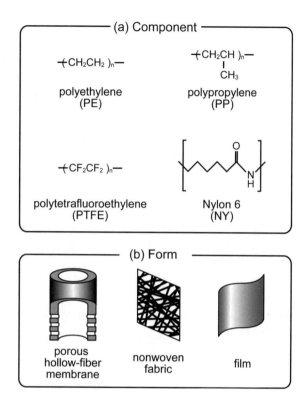

(a) Component

$+CH_2CH_2$ $)_n$—

polyethylene
(PE)

$+CH_2CH$ $)_n$—
|
CH_3

polypropylene
(PP)

$+CF_2CF_2$ $)_n$—

polytetrafluoroethylene
(PTFE)

Nylon 6
(NY)

(b) Form

porous
hollow-fiber
membrane

nonwoven
fabric

film

the pores [3–6]. Furthermore, high-capacity binding of proteins proceeds through multilayer binding of proteins into the extended graft chain [7].

1.1.2   Classification of Radiation-Induced Graft Polymerization

Electron-beam or gamma-ray irradiation generates radicals on/in trunk polymers. The contact of the irradiated trunk polymer with a vinyl monomer initiates graft polymerization. Radiation-induced graft polymerization techniques are divided into two types in terms of method of irradiation: preirradiation and simultaneous grafting. As to the former method, the trunk polymer is irradiated before a vinyl monomer is grafted, whereas in the latter method, a mixture of the trunk polymer and vinyl monomer is irradiated to simultaneously induce radical generation and grafting.

We have adopted preirradiation graft polymerization for two reasons: (1) The irradiation process is separated from the grafting process, and (2) the formation of homopolymers is suppressed. When the vinyl polymer is relatively volatile, the graft polymerization is carried out in the gas phase, as illustrated in Fig. 1.3a. A monomer is almost completely consumed by graft polymerization in the gas phase; however, the distribution of the amount of the polymer grafted onto the trunk polymer results from the partial distribution of vapor pressure in the reactor. In contrast, graft polymerization in the liquid phase (Fig. 1.3b) yields the amount of graft chain uniformly over the entire volume as a result of stirring or convection of a vinyl monomer solution. The degree of grafting, defined in Eq. 1.1, can be adjusted by controlling the reaction temperature and reaction time. We have employed reaction temperatures ranging from 4 to 60 °C.

$$\text{Degree of grafting} = 100(W_1 - W_0)/W_0, \qquad (1.1)$$

where W_0 and W_1 are the masses of the trunk polymer and grafted polymer, respectively. A 100% degree of grafting means that the mass of the resulting polymer increased by a factor of two; that is, the mass of the trunk polymer was

Fig. 1.3 Gas- and liquid-phase graft polymerization on the laboratory scale

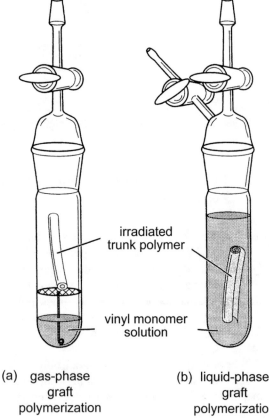

irradiated
trunk polymer

vinyl monomer
solution

(a) gas-phase
graft
polymerization

(b) liquid-phase
graft
polymerization

equivalent to that of the graft chain. Also, the formation site of the graft chain varies with the solvent used for the vinyl monomer and its concentration.

1.1.3   Storage of Radicals

Specifications of the radicals formed on the trunk polymers, such as species, location, and lifetime, are required to tailor the adsorbents by radiation-induced graft polymerization. For example, electron-beam irradiation of polyethylene generates three types of radical, i.e., alkyl, allyl, and peroxy radicals. The decay of each type of radical in air was compared with that in nitrogen atmosphere at various temperatures.

The concentrations of allyl and peroxy radicals were constant in either atmosphere. In contrast, a higher concentration of alkyl radicals was found in the nitrogen atmosphere than in air at the same temperature after irradiation. This difference can be explained by the accessibility of oxygen to the radicals. Because the vinyl monomer also diffuses into oxygen-accessible sites, the radical contributing to graft polymerization will be an alkyl radical instead of an allyl radical.

As shown in Fig. 1.4, after being maintained for 150 h in air at a dry-ice temperature of 195 K, almost no decrease in alkyl radical concentration was observed [8]. This suggests that the radical is maintained in dry ice where the matrix polymers have no mobility below a glass transition temperature of 203 K (−70 °C) for polyethylene. On the other hand, at an ambient temperature of 298 K (25 °C), approximately the initial concentration of alkyl radicals decreased by 90% during 10 h of contact with air.

A radical attracts hydrogen in a neighboring group, and the radical replaces the hydrogen. This phenomenon is regarded as the diffusion of radicals in the polymer. Uezu et al. [8] analyzed radical decay at various temperatures in air and nitrogen

Fig. 1.4 Alkyl radical decay at various storage temperatures

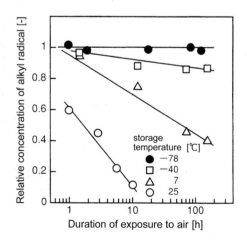

atmosphere on the basis of the radical diffusion model. When the polymer network is frozen, hydrogen replacement is inhibited; therefore, no radical decay is observed.

This absence of radical decay, namely capability of radical storage, has an industrial significance: The process of graft polymerization is separable from the process of the irradiation with electron beams or gamma rays. The trunk polymer was irradiated with an electron beam not in the laboratory of Chiba University but at a commercial irradiation facility located at Tsukuba City, approximately 100 km from Chiba. The irradiated trunk polymer was transported by vehicle to Chiba University at a low temperature maintained with dry ice. As long as the irradiated trunk polymer is freeze-dried, the radicals may be preserved, even in air.

1.1.4   Irradiation Facilities

Radiation-induced graft polymerization is applicable to existing polymeric materials of variable forms. In this book, porous hollow-fiber membranes, porous sheets, fibers, nonwoven fabrics, films, and particles are the forms adopted for trunk polymers for grafting. With the accelerator shown in Fig. 1.5, electrons emitted from an electron gun are accelerated at a prescribed electrical voltage and current. Accelerated electrons in a high-vacuum chamber are deflected by a magnetic field and then penetrate through a tungsten-foil window to attack the trunk polymer that is placed beneath the window. On a laboratory scale, the trunk polymer packed into a nitrogen-filled plastic bag on a conveyor is irradiated at a prescribed dose.

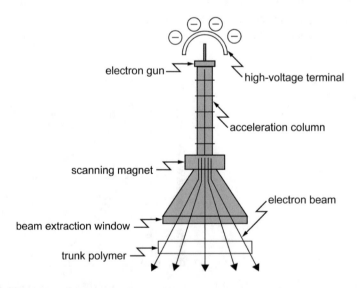

Fig. 1.5 Electron accelerator

A cobalt-60 source was used as a gamma-ray source to produce the radicals on the trunk polymer. An electron beam is temporarily emitted using electrical power, whereas gamma rays are continuously emitted from a Co-60 source stored deep in the water of a pool. Cerenkov radiation is observed in the water near the source. A panel consisting of Co-60 sources is transported with elevators to an irradiation room, as illustrated in Fig. 1.6. The gamma rays irradiate onto the trunk polymer placed in the room. The dose can be adjusted by changing the distance between the panel and the trunk polymer and the irradiation time. After irradiation, the panel is returned to the original site in the water pool.

Currently, irradiation from an electron beam or gamma rays is primarily applied to the sterilization of various materials at a highly precise dose to ensure acceptable cost performance. This state-of-the-art method is preferred for radiation-induced graft polymerization because the irradiation is carried out on the industrial scale to produce radicals on the trunk polymers.

1.1.5   Functional Groups and Ligands

The form and component of the trunk polymer should be selected in accordance with the ultimate application of the grafted materials. Depending on the target ions and molecules, a specific functional group should be adopted. Using optimal combinations of the trunk polymer, vinyl monomer, and functional group in radiation-induced graft polymerization is advantageous over other modifications. Polyethylene, polypropylene, and nylon 6 are all economically feasible compared

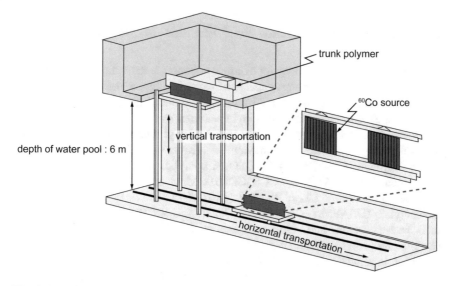

Fig. 1.6 Facility for gamma irradiation

with polytetrafluoroethylene. Polyimide and polyethylene terephthalate are not reactive in graft polymerization because they form stable resonance structures involving electrons after irradiation.

The targets in the separation technique determine suitable chemical moieties or functional groups. Mostly, the targets are ions dissolved in aqueous media: Heavy metal ions dissolved in ultrapure water at extremely low concentrations or rare metal ions dissolved in seawater, e.g., ionic uranium species at a concentration of 3 ppb or 3 μg/L dissolved in seawater. The concentration ratio of target ions to foreign ions sometimes ranges over ten million.

The functional groups introduced into the graft chain must be selected according to the target ions and molecules to be captured and immobilized. Previous studies on separation and purification of various substances have provided feasible candidate functional groups. For example, an iminodiacetate group [9, 10] as a chelate-forming group is selected for the removal of trace amounts of heavy metal ions such as lead and copper ions from water. A sulfonic acid group [11, 12] as a cation-exchange group is selected for the recovery of lysozyme from egg-white solution.

Interactions between target ions and functional groups are classified into the following: ion exchange or electrostatic, chelate forming, hydrophobic, affinity, and covalent bonding. Functional groups are introduced into the polymer chain and are grafted on trunk polymers of arbitrary forms. In addition, instead of functional groups and ligands, extractants and inorganic compounds are impregnated onto the graft chain. In contrast, to prevent the nonselective or irreversible adsorption of proteins onto the graft chain, hydrophilic groups such as a diol group [13] and ampholytes [14] such as sulfobetaine can be introduced into the graft chain.

1.1.6 Preparation Schemes for Polymeric Adsorbents

We select the form and component of trunk polymers suitable for specific applications and the functional groups capable of capturing targets before we design the preparation scheme for adsorbents by radiation-induced graft polymerization and subsequent chemical modification. Vinyl monomers used for radiation-induced graft polymerization include functional and reactive vinyl monomers and crosslinkers. The following functional vinyl monomers are commercially available: sodium styrene sulfonate (SSS, $CH_2=CHC_6H_4SO_3Na$), acrylic acid (AAc, $CH_2=CHCOOH$), methacrylic acid (MAc, $CH_2=CCH_3COOH$), vinyl benzyl trimethyl ammonium chloride (VBTAC, $CH_2=CHC_6H_4CH_2N(CH_3)_3Cl$), and diethylaminoethyl methacrylate (DEAEMA, $CH_2=CCH_3COOCH_2CH_2N(CH_3)_2$). These vinyl monomers possess cation- and anion-exchange groups, i.e., negatively and positively charged groups, respectively.

Direct introduction of functional groups with functional vinyl monomers is favorable. For example, cation- and anion-exchange fibers are readily prepared by grafting SSS and VBTAC onto irradiated nylon fibers, respectively, as shown in

Fig. 1.7. However, hydrophilic vinyl monomers exhibit low accessibility to the irradiated hydrophobic polyethylene as a trunk polymer, resulting in a negligibly low grafting rate. To improve accessibility, the cografting of neutral (uncharged) vinyl monomers such as 2-hydroxyethyl methacrylate (HEMA, $CH_2=CCH_3COOCH_2$ CH_2OH) and the hydrophilic vinyl monomers was effective [15]. A possible mechanism of the cografting of HEMA and SSS is illustrated in Fig. 1.8.

A representative of reactive vinyl monomers is an epoxy-group-containing vinyl monomer, glycidyl methacrylate (GMA, $CH_2=CCH_3COOCH_2CHOCH_2$). The epoxy group of GMA readily reacts with various chemical moieties such as amino and thiol groups (Fig. 1.9). On the basis of the GMA-divinylbenzene copolymer, Svec's research group [16–18] prepared a number of functional polymeric beads for the collection of ions, the purification of proteins, and the removal of gases. We referred to their work to introduce functional groups into the poly-GMA chain grafted by radiation-induced graft polymerization. In our studies, the molecular masses of the chemicals that ring-open the epoxy group of the poly-GMA chain range from 17 for ammonia [19] to approximately 170,000 for an antigen [20]. Increasing the molecular mass of a chemical decreases the functional group density because of accessibility to the epoxy group.

GMA, a representative reactive monomer, is also referred to as a precursor monomer. GMA is one of the most versatile and convenient vinyl monomers in functionalizations. In its favor, the ester group contained in the molecular structure of GMA is resistant to hydrolysis within a wide range of pHs at ambient temperature.

Crosslinkers such as divinylbenzene (DVB) [21] and ethyleneglycol dimethacrylate (EGDMA) [22, 23] were used to crosslink graft chains. Crosslinking was used to suppress the extension of the charged graft chain and to decrease the mobility of the graft chain. For example, the liquid permeability of a charged

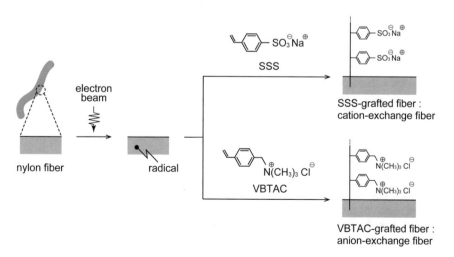

Fig. 1.7 Preparation schemes for ion-exchange fibers

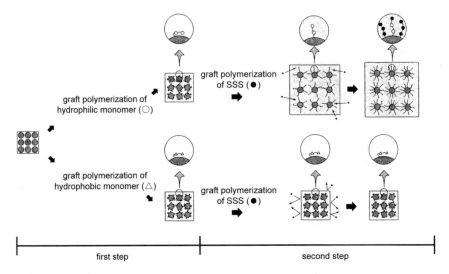

Fig. 1.8 Mechanism of two-step graft polymerization of sodium styrene sulfonate (SSS) and hydrophilic monomer Reprinted with permission from Ref. [15]. Copyright 1993 American Chemical Society

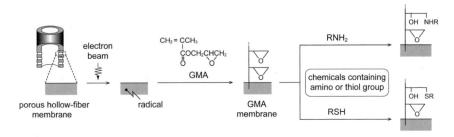

Fig. 1.9 Introduction of amino- or thiol-group-containing reagent into epoxy group of graft chain

graft-chain-immobilized porous hollow-fiber membrane was controllable by cografting of DVB and GMA.

1.1.7 Degree of Grafting

The quantity as well as quality of immobilized functional groups governs the performance of polymeric adsorbents. The measure of quantity is functional group density, which is directly related to the binding capacity for targets. For example, the binding capacity of an anion-exchange adsorbent for phosphate ions as low molecular mass ions increases linearly with anion-exchange-group density [24], as shown in Fig. 1.10a, where the abscissa in this figure, molar conversion of epoxy groups to trimethylammonium groups, is equivalent to the anion-exchange-group density. The binding capacity for albumin as a high molecular mass ion increased with anion-exchange-group density [24] (Fig. 1.10b). This increase is caused by the extension of the graft chain owing to mutual electrostatic repulsion among the graft chains: The extension of the graft chain provides the 3D binding space for the protein. In contrast, the binding capacity of the graft chain containing hydrophobic and affinity ligands for proteins is determined not by ligand density but by the specific surface area.

Our goal is to develop polymeric adsorbents for practical use with binding capacity and rate that meet the needs of the target application; therefore, radiation-induced grafting should modify not only the surface of the trunk polymer but also its bulk. Ordinarily, the degree of grafting is prescribed to be from 50 to 150% as opposed to 1–5%. Although grafting up to 200–300% is possible, an increased amount of the graft polymer penetrating the matrix of the trunk polymer can decrease the physical strength of the polymer.

1.1.8 Structure of Graft-Type Polymeric Adsorbents

As illustrated in Fig. 1.11a, from the term "graft," the graft chain is depicted as a branch growing from a tree trunk; however, this imagery is not accurate. Figure 1.11b is closer to an actual condition. Let us describe the structure of a grafted polymeric material prepared by preirradiation graft polymerization of GMA onto a high-density polyethylene membrane of a hollow fiber. First, high-density polyethylene consists of crystalline and amorphous domains. The crystalline

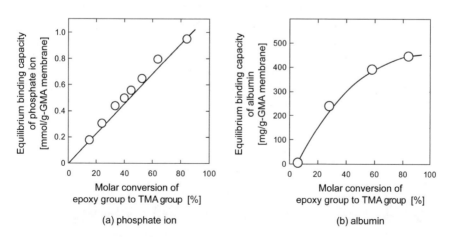

Fig. 1.10 Equilibrium binding capacity of anion versus molar conversion of epoxy groups to trimethylammonium (TMA) groups

domain is an integrated structure of lamella formed by the tightly folded poly-ethylene chain, whereas the amorphous domain is formed by loose polyethylene chain.

This hollow fiber has inner and outer diameters of 2 and 3 mm, respectively. To convert a membrane used for microfiltration into a polymeric adsorbent, we pro-duce radicals on the entire polyethylene matrix irrespective of the crystalline and amorphous domains by irradiation with an electron beam or gamma rays. The irradiation depth can be adjusted by regulating the energy applied to the accelerator or the prescribed distance from the Co-60 source.

1.1.9 Polymer Brush and Polymer Root

When a porous polymeric material such as a porous hollow-fiber membrane or a sheet is irradiated with an electron beam or gamma rays at a dose sufficient for electrons to penetrate the porous polymer, radicals are formed uniformly throughout its entire volume. For example, a porous polyethylene sheet consists of the crys-talline and amorphous domains as a matrix. Neither an electron beam nor gamma rays can distinguish this structural difference.

Polymer chains grafted onto a porous polyethylene sheet can be classified into two categories: polymer chains extending from the pore surface toward the pore interior and polymer chains invading the polymer matrix. The former is referred to as a polymer brush and the latter as a polymer root. The polymer brush produced by various graft polymerization techniques, e.g., plasma-induced graft polymerization and ATRP, has been studied over the past two decades. The polymer root is char-acteristic of a polymer chain produced by radiation-induced graft polymerization.

The polymer brush and polymer root in porous polymeric adsorbents share respective roles in binding proteins during their permeation. The polymer brush captures proteins during the permeation of a protein solution through the pores that are rimmed by the polymer brush, whereas the polymer root swells the polymer matrix, which minimizes the pore size reduction caused by the polymer brush.

The mole percentages of the polymer brush and polymer root in the graft chain have not been elucidated because it is difficult to isolate the polymer brush from the polymer root. Uchiyama et al. [25] proposed a novel method for determining the

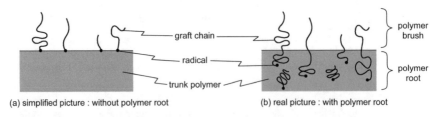

(a) simplified picture : without polymer root   (b) real picture : with polymer root

Fig. 1.11 Graft chain consisting of polymer brush and polymer root

mole percentage of the polymer brush on the basis of the swelling behavior accompanied by the introduction of charged groups into the graft chain with water as a solvent. This method is detailed in Chap. 2.

The ratio of polymer brush to polymer root varies according to the component of the trunk polymer, the kind of vinyl monomer and solvent, and grafting conditions. Relatively low molecular mass ions can easily access the polymer root as well as the polymer brush. To prepare ion-exchange membranes from polymeric films such as polyethylene and nylon films, the polymer root must penetrate the amorphous domain across the film thickness before the ion-exchange groups are introduced into the polymer root. The polymer root plays an essential role in governing the performance of ion-exchange membranes.

1.1.10   Molar Conversion

The preparation schemes for typical polymeric ion exchangers by radiation-induced graft polymerization are shown in Fig. 1.12. These schemes using GMA as a precursor monomer consist of three steps: (1) irradiation by an electron beam or gamma rays to produce radicals on the trunk polymer, (2) graft polymerization of GMA, and (3) conversion of the epoxy group produced into an ion-exchange group with reagents such as amines. In this figure, cation- and anion-exchange groups and ampholytes are immobilized on the poly-GMA chain grafted onto a porous hollow-fiber membrane.

Here, the molar conversion of the epoxy group of the poly-GMA graft chain into functionalities or the degree of functionalization is defined as

molar conversion of epoxy group into functionality [%]

$$= 100(\text{moles of functionality introduced})/(\text{moles of epoxy group before modification})$$

$$= 100\left[(W_2 - W_1)/M_{HR}\right] / \left[(W_1 - W_0)/142\right],$$

$$(1.2)$$

where W_2 is the mass of the polymer after the introduction of the functional group by the addition of a chemical reagent HR to the epoxy group. M_{HR} and the value 142 are the molecular masses of HR and GMA, respectively.

The epoxy-ring opening of the graft chain with HR is convenient in that neither side reaction nor successive reaction occurs during the modification. Therefore, the molar conversion is readily evaluated from the mass gain of the polymer. An optimum molar conversion may be determined in terms of the physical strength of the adsorbent as a product. When a complete or 100% molar conversion is not attained or allowed, the remaining epoxy group of the graft chain remains unreacted or is converted into another functional group. For example, for the collection of a desirable protein, some epoxy groups are converted into diethylamino groups by

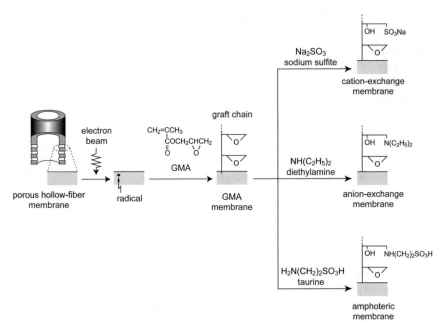

Fig. 1.12 Introduction of ion-exchange groups into graft chains

reaction with diethylamine to selectively capture the desired protein, and the remaining epoxy groups are reacted with 2-hydroxyethylamine to retard the capture of undesirable proteins [19].

1.2   Evaluation of Performance in Separation

1.2.1   Performance of Adsorbents

Separation methods include distillation, absorption, extraction, adsorption, and ion exchange. The polymeric materials for separation described in this book are the adsorbents that contain the functional groups introduced into the graft chain via covalent bonding. In addition, we describe the polymeric materials designed for solid-phase extraction: Various extractants are impregnated onto the graft chain via hydrophobic interactions. Intrinsically, these materials are classified as adsorbents. This book also deals with enzyme-immobilized polymeric materials in which enzymes are linked to the graft chain via electrostatic interactions, and subsequent crosslinking between enzymes. Terminologically, the difference between impregnation and immobilization is not clear; however, here we use impregnation for extractants and immobilization for enzymes.

Evaluation of the performance of the adsorbents prepared by radiation-induced graft polymerization is necessary to quantitatively demonstrate their advantages over conventional adsorbents. Essential criteria for adsorbents are high adsorption rate, binding capacity, and durability for repeated use of adsorption and elution. Furthermore, to consider the commercialization of an adsorbent, the product cost should also be considered.

1.2.2   Flow-Through Mode

Adsorbents are evaluated either in the flow-through mode or in the batch mode. In the flow-through mode, a column is employed, into which adsorbent beads are packed in most cases.

We modified a polyethylene porous hollow-fiber membrane to append various functionalities for recovering valuable ions and molecules and removing undesirable ions and molecules. The binding performance of modified porous hollow-fiber membranes is evaluated in the permeation mode by using the experimental apparatus shown in Fig. 1.13. Both the length of the hollow fiber effective for liquid permeation and the initial concentration of the feed may vary; therefore, to reasonably compare the breakthrough curves, a dimensionless treatment of experimental data is convenient. The values along the y-axis of the breakthrough curve are converted to the concentration ratio of the effluent to the feed, whereas the values along the x-axis are converted to dimensionless effluent volume by dividing the effluent volume by the hollow-fiber membrane volume. Here, the hollow-fiber membrane volume excluding the lumen is defined as

$$\text{membrane volume} = (1/4)\pi\left(d_o^2 - d_i^2\right)L, \qquad (1.3)$$

where d_o, d_i, and L are the outer and inner diameters and the effective length of the hollow fiber, respectively. The breakthrough curve showing dimensionless effluent concentration versus dimensionless effluent volume is referred to as the dimensionless breakthrough curve (Fig. 1.14).

1.2.3   Equilibrium and Dynamic Binding Capacities

The breakthrough curves frequently assume a sigmoid shape. Initially, the effluent concentration of the target ion or molecule ion remains zero. As the effluent volume increases, the effluent concentration gradually increases and starts to level off. Finally, the effluent concentration reaches the feed concentration, which means that an equilibrium is attained. In contrast, an original porous hollow-fiber membrane exhibits almost no breakthrough behavior; only the liquid previously filling the pore is excluded by the feed.

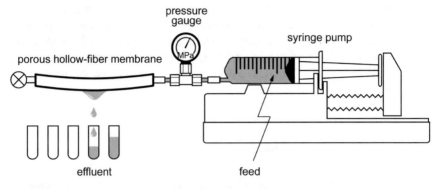

Fig. 1.13 Experimental apparatus for the determination of breakthrough curves of porous hollow-fiber membrane at various flow rates and feed concentrations

Fig. 1.14 Dimensionless breakthrough and elution curves of diethylamino-type anion-exchange porous hollow-fiber membrane for BSA at various flow rates Reprinted with permission from Ref. [27]. Copyright 1995 Elsevier

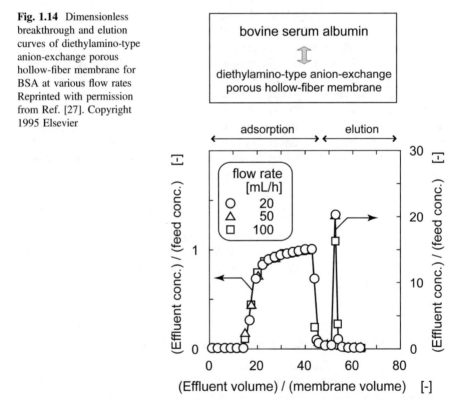

The amount of target ions adsorbed onto the adsorbent is evaluated by integrating the concentration differences between the feed and the effluent. After equilibration, the amount of target ions adsorbed per mass of adsorbent is referred to as the equilibrium binding capacity (EBC). EBC is evaluated as follows:

$$\text{EBC} = \int_0^{V_e} (C_0 - C)\mathrm{d}V / W, \tag{1.4}$$

where C_0 and C are the concentrations of the feed and effluent, respectively. V and V_e are the effluent volume and the effluent volume when the concentration of the feed reaches that of the effluent, respectively. W is the mass of the adsorbent. The numerator in this equation is equivalent to the shaded area in Fig. 1.15.

EBC is a static property of the adsorbent. For the recovery of valuable targets such as precious metal ions and pharmaceutical proteins, permeation is stopped before the target is detected because the leakage of the targets is wasteful. On the other hand, for the removal of dangerous targets such as heavy metal ions and undesirable proteins, the permeation is stopped because the leakage of the targets is unfavorable.

To quantify the breakthrough curve, the breakthrough point is defined as the effluent volume at which the effluent concentration of the target ions or molecules reaches 10% of the feed concentration. Instead of 10%, 5% is often adopted as the breakthrough point. The correct determination of the effluent volume at which the target starts to leak at the column exit or the outside surface of the hollow fiber is experimentally difficult because the detection sensitivity of the target depends on analytical instruments and procedures. A breakthrough point of 10 or 5% easily interpolated from the breakthrough curve is convenient for chemical engineers.

The amount of target adsorbed per mass of adsorbent up to the breakthrough point is referred to as the dynamic binding capacity (DBC). DBC is also referred to as practical binding capacity. For practical use, DBC is much more significant than EBC. DBC is a dynamic property that depends on the flow rate of the solution or the residence time of the target across the column or through the porous hollow-fiber membrane.

Fig. 1.15 Determination of breakthrough curve

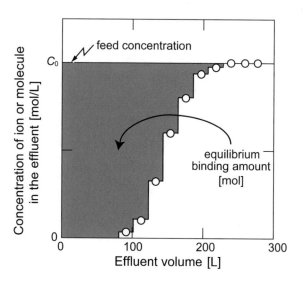

1.2.4 Capturing Efficiency

As the flow rate of the feed to the bead-packed column increases, the breakthrough point shifts to the left and the breakthrough curve expands (Fig. 1.16) because the time required for the target to diffuse into the interior of the bead is insufficient. In contrast, as the flow rate decreases, the breakthrough point shifts to the right. This is due to a sufficient time for the diffusion of the target into the interior of the bead. According to the flow rate of the feed, the shape of the breakthrough curve changes while retaining EBC because EBC is a static property of the beads packed into the column.

The dependence of the shape of the breakthrough curve on the flow rate in the permeation mode reflects the rate-determining step in the overall adsorption process. The observation that the breakthrough curves overlapped regardless of the flow rate of the target-containing solution or residence time demonstrates negligible diffusional mass-transfer resistance of the targets and instantaneous ion-exchange or chelate-forming reactions (Fig. 1.14) [26]. This characteristic is favorable for practical use: The same amount of target is captured from the same volume of solution at any flow rate. This characteristic is also ideal for the adsorption operation because the higher the flow rate, the higher the overall adsorption rate.

1.2.5 Batch Mode

In the batch mode, a plastic container is used in place of the column. The concentration decay or concentration change over time of target ion or molecule is measured at a prescribed mass ratio of liquid to adsorbent. The liquid in the plastic container is shaken or stirred at a constant temperature. At appropriate intervals,

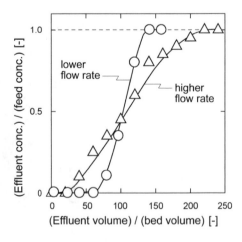

Fig. 1.16 Dependence of breakthrough curves of adsorbent beads packed in a bed on flow rate of target solution

an aliquot of the liquid was sampled to determine the concentration of the target, and the amount of the target adsorbed is evaluated using the following mass balance equation:

$$q = V(C_i - C_t)/W, \tag{1.5}$$

where C_i and C_t are the initial concentration and the concentration at the sampling time, respectively, and V and W are the liquid volume and the mass of the adsorbent, respectively. When the concentration remains unchanged, an equilibrium is considered to be attained. Thus, one set of experimental data for an adsorption isotherm is obtained. If the concentration decay is observed even at a high-mass ratio of liquid to adsorbent, the adsorbent exhibits a high equilibrium binding capacity.

1.2.6 Repeated Cycles of Adsorption and Elution

To recover proteins and rare metal ions and to reuse the adsorbents, elution of targets adsorbed onto the adsorbents with an appropriate eluent is essential. First, a strongly acidic cation-exchange resin containing the sulfonic acid group ($-SO_3H$) is conditioned with sodium hydroxide solution to convert an H-type resin to an Na-type resin. Second, after washing with deionized water, target cations such as calcium and magnesium ions are captured by $-SO_3Na$ via cation exchange. Third, bound target cations are quantitatively eluted with hydrochloric acid and then washed with deionized water. To reuse the resin, the resin is conditioned with sodium hydroxide solution again. One cycle of adsorption and elution consists of a series of conditioning, washing, adsorption, washing, elution, and washing.

As shown in Fig. 1.14, the breakthrough curve of an anion-exchange porous hollow-fiber membrane for albumin is completed before the elution curve is determined. From these curves, elution percentage, defined as the percentage of the amount eluted to the amount adsorbed, is evaluated. An elution percentage of 100% or a quantitative elution reproduces the breakthrough curve [27].

We use GMA as a precursor monomer, and GMA is graft-polymerized onto trunk polymers for the introduction of the epoxy group into various functionalities. The ester group of the poly-GMA chain did not hydrolyze during elution at a low pH and ambient temperature.

1.3 Innovative Materials for Separation

The famous Austrian economist Shumpeter (1883–1950) defined innovation as a new combination. The polymeric adsorbents for separation and reaction described in this book are new combinations of existing polymers with a graft polymer prepared by radiation-induced graft polymerization. Therefore, the novel polymeric adsorbents we have developed are innovative products. Over thirty years, we have tried to develop graft-type adsorbents with higher performance than conventional adsorbents.

Three achievements and their significance are summarized as follows:

(1) Advantage at the macroscopic level: The use of suitable porous and nonporous polymeric materials selected from the existing ones as trunk polymers facilitates adsorption.

To improve the performance of adsorbents in separation, we select functional groups or ligands suitable for capturing target ions and molecules and forms or porous structures useful for minimizing mass-transfer resistance. The selection of suitable functional groups or ligands is derived from voluminous findings of previous studies. The selection of useful forms or porous structure is based on consideration of two transport modes: diffusion driven by the concentration gradient and convection assisted by the bulk flow.

Minimization of the diffusional mass-transfer path of ions or molecules by convective transport is effective for promoting overall mass transfer. For example, an ion-exchange porous hollow-fiber membrane enables target ions to be transported by permeative flow of the solution driven by transmembrane pressure; target ions are transported to the vicinity of the ion-exchange graft chain that is appended onto the pore surface uniformly across the membrane. Moreover, ion-exchange polymer-brush-immobilized particles reduce the diffusional mass-transfer path of ions along the polymer-brush length and not across the particle radius.

(2) Advantage at the microscopic level: Control of polymer brush and polymer root produces novel and sophisticated adsorbents.

Thus far, graft polymerization has been employed for surface modification. However, we extended application of radiation-induced graft polymerization to bulk modifications to prepare high-performance polymeric adsorbents. First, to achieve higher binding capacity of graft-type adsorbents than of conventional adsorbents, a considerable amount of vinyl monomer is graft-polymerized onto the trunk polymer that was previously irradiated entirely over its entirety.

In this book, the classification of graft chains by formation site is proposed to explain the functions of the graft chains. For example, when an ion-exchange graft chain is appended onto a polyethylene porous membrane with the hollow-fiber form, polymer roots invading the polyethylene matrix swell the entire membrane volume, which compensates for the reduction in the volume caused by the polymer brush extending toward the pore interior from the pore surface.

Smaller targets such as metal ions bind to both the polymer brush and polymer root, whereas larger targets such as proteins bind to only the polymer brush in multilayers via multipoints. Control of the mass ratio of polymer brush to polymer root by varying preparation conditions will enable the preparation of novel and sophisticated polymeric adsorbents.

(3) Practical advantage: The separation of the graft polymerization process from the irradiation process enhances the mass production of grafted materials.

The fact that the radicals do not decay as long as the irradiated trunk polymer is stored below its glass transition temperature was experimentally demonstrated. This has led to the separation of the grafting process from the irradiation process that some companies specialize in. These advantages enable the mass production of high-performance adsorbents with new forms. In the following chapters, we describe a new-found science involving the graft chain and newly developed applications of the graft chain.

References

1. K. Saito, T. Sugo, High-performance polymeric materials for separation and reaction, prepared by radiation-induced graft polymerization. in *Radiation Chemistry: Present Status and Future Trends*, eds. by C.D. Jonah, M. Rao (Elsevier, 2001), pp. 671–704
2. S. Sugiyama, S. Tsuneda, K. Saito, S. Furusaki, T. Sugo, K. Makuuchi, Attachment of sulfonic acid groups to various shapes of PE, PP and PTFE by radiation-induced graft polymerization. React. Polym. **21**, 187–191 (1993)
3. K. Saito, S. Tsuneda, M. Kim, N. Kubota, K. Sugita, T. Sugo, Radiation-induced graft polymerization is the key to develop high-performance functional materials for protein purification. Radiat. Phys. Chem. **54**, 517–525 (1999)
4. K. Saito, Charged polymer brush grafted onto porous hollow-fiber membrane improves separation and reaction in biotechnology. Sep. Sci. Technol. **37**, 535–554 (2002)
5. T. Kawai, K. Saito, W. Lee, Protein binding to polymer brush, based on ion-exchange, hydrophobic, and affinity interactions. J. Chromatogr. B **790**, 131–142 (2003)
6. K. Saito, Preparation of porous adsorbers and supports most favorable for separation by using radiation-induced graft polymerization. Kobunshi Ronbunshu **71**, 302–312 (2014)
7. S. Matoba, S. Tsuneda, K. Saito, T. Sugo, Highly efficient enzyme recovery using a porous membrane with immobilized tentacle polymer chains. Nat. Biotechnol. **13**, 795–797 (1995)
8. K. Uezu, K. Saito, S. Furusaki, T. Sugo, I. Ishigaki, Radicals contributing to preirradiation graft polymerization onto porous polyethylene. Radiat. Phy. Chem. **40**, 31–36 (1992)
9. S. Tsuneda, K. Saito, S. Furusaki, T. Sugo, J. Okamoto, Metal collection using chelating hollow-fiber membrane. J. Membr. Sci. **58**, 221–234 (1991)
10. H. Yamagishi, K. Saito, S. Furusaki, T. Sugo, I. Ishigaki, Introduction of a high-density chelating group into a porous membrane without lowering the flux. Ind. Eng. Chem. Res. **30**, 2234–2237 (1991)
11. H. Shinano, S. Tsuneda, K. Saito, S. Furusaki, T. Sugo, Ion exchange of lysozyme during permeation across a microporous sulfopropyl-group-containing hollow fiber. Biotechnol. Prog. **9**, 193–198 (1993)
12. S. Tsuneda, H. Shinano, K. Saito, S. Furusaki, T. Sugo, Binding of lysozyme onto a cation-exchange microporous membrane containing tentacle-type grafted polymer branches. Biotechnol. Prog. **10**, 76–81 (1994)
13. M. Kim, J. Kojima, K. Saito, S. Furusaki, T. Sugo, Reduction of nonselective adsorption of proteins by hydrophilization of microfiltration membranes by radiation-induced grafting. Biotechnol. Prog. **10**, 114–120 (1994)
14. S. Matsuno, K. Iwanade, D. Umeno, K. Saito, H. Ito, M. Sakamoto, Carboxybetaine-group immobilized onto pore surface reduced protein adsorption to porous membrane. Membrane (Maku) **35**, 86–92 (2010)

15. S. Tsuneda, K. Saito, S. Furusaki, T. Sugo, K. Makuuchi, Simple introduction of sulfonic acid group onto polyethylene by radiation-induced cografting of sodium styrenesulfonate with hydrophilic monomers. Ind. Eng. Chem. Res. **32**, 1464–1470 (1993)
16. J. Kalal, F. Svec, V. Marousek, Reactions of epoxide groups of glycidyl methacrylate copolymers. J. Polym. Sci. **47**, 155–166 (1974)
17. H. Hrudkova, F. Svec, J. Kalal, Reactive polymers. XIV. Hydrolysis of the epoxide groups of copolymer glycidyl methacrylate-ethylene dimethacrylate. Br. Polym. J. **9**, 238–240 (1977)
18. T.B. Tennikova, M. Bleha, F. Svec, T.V. Almazova, B.G. Belenkii, High-performance membrane chromatography of proteins. A novel method of protein separation. J. Chromatogr. **555**, 97–107 (1991)
19. I. Koguma, K. Sugita, K. Saito, T. Sugo, Multilayer binding of proteins to polymer chains grafted onto porous hollow-fiber membranes containing different anion-exchange groups. Biotechnol. Prog. **16**, 456–461 (2000)
20. S. Nishiyama, A. Goto, K. Saito, K. Sugita, M. Tamada, T. Sugo, T. Funami, Y. Goda, S. Fujimoto, Concentration of 17β-estradiol using an immunoaffinity porous hollow-fiber membrane. Anal. Chem. **74**, 4933–4936 (2002)
21. K. Saito, M. Ito, H. Yamagishi, S. Furusaki, T. Sugo, J. Okamoto, Novel hollow-fiber membrane for the removal of metal ion during permeation: preparation by radiation-induced cografting of a cross-linking agent with reactive monomer. Ind. Eng. Chem. Res. **28**, 1808–1812 (1989)
22. K. Saito, K. Saito, K. Sugita, M. Tamada, T. Sugo, Cation-exchange porous hollow-fiber membrane prepared by radiation-induced cografting of GMA and EDMA which improved pure water permeability and sodium ion adsorptivity. Ind. Eng. Chem. Res. **41**, 5686–5691 (2002)
23. G. Wada, R. Ishihara, K. Miyoshi, D. Umeno, K. Saito, S. Asai, S. Yamada, H. Hirota, Crosslinked-chelating porous sheet with high dynamic binding capacity of metal ions. Solv. Extr. Ion Exch. **31**, 210–220 (2013)
24. R. Shibahara, K. Hagiwara, D. Umeno. K. Saito, T. Sugo, Preparation of size-exclusion polymer chain grafted onto the pore surface of a porous hollow-fiber membrane. Membrane (Maku) **34**, 220–226 (2009)
25. S. Uchiyama, R. Ishihara, D. Umeno, K. Saito, S. Yamada, H. Hirota, S. Asai, Determination of mole percentages of brush and root of polymer chain grafted onto porous sheet. J. Chem. Eng. Japan **46**, 414–419 (2013)
26. S. Konishi, K. Saito, S. Furusaki, T. Sugo, Sorption kinetics of cobalt in chelating porous membrane. Ind. Eng. Chem. Res. **31**, 2722–2727 (1992)
27. S. Tsuneda, K. Saito, S. Furusaki, T. Sugo, High-throughput processing of protein using a porous and tentacle anion-exchange membrane. J. Chromatogr. A **689**, 211–218 (1995)

Chapter 2
Scientific Findings on Graft Chains

Abstract A graft chain immobilized onto a trunk polymer by radiation-induced graft polymerization has a free end and an immobile end. Depending on the formation site, the graft chain is divided into a polymer brush extending from the surface of the trunk polymer and a polymer root entering the matrix of the trunk polymer. The graft chain will extend or shrink depending on the density of the charged group of the graft chain and the ionic strength of the liquid surrounding the graft chain. An extended polymer brush captures proteins in multilayers via multipoints. When a graft chain is immobilized over a porous membrane, the permeability of the liquid through the porous membrane reflects the static and dynamic behavior of the graft chain. Also, the graft-chain phase diffusion of metal ions and proteins occurs, driven by the gradient of the number of ions and proteins bound by the graft chain.

Keywords Polymer brush · Polymer root · Multilayering of protein
Graft-chain phase diffusion

Irradiation of a porous polymer with an electron beam or gamma rays with a relatively high energy produces radicals throughout the matrix of the porous polymer. Consequently, when a vinyl monomer is accessible to a radical, a polymer chain grows from the radical to form a graft chain. Some graft chains growing from radicals located deep within the matrix penetrate the pore surface toward the pore interior. Other graft chains remain in the polymer matrix. Of the graft chains prepared by radiation-induced graft polymerization, the part extending from the pore surface toward the pore interior and that invading the matrix is referred to as the polymer brush and polymer root, respectively (Fig. 1.11). A polymer brush can capture high-mass molecules such as proteins, whereas a polymer root allows the matrix to swell. The swelling of the matrix caused by a polymer root can minimize the reduction in pore size caused by a polymer brush.

Most graft chains have both free and immobilized ends. The "graft-chain phase," which consists of graft chains, behaves like a liquid. For example, when charged groups are introduced into graft chains, the charged graft-chain phase is regarded as a

© Springer Nature Singapore Pte Ltd. 2018
K. Saito et al., *Innovative Polymeric Adsorbents*,
https://doi.org/10.1007/978-981-10-8563-5_2

polyelectrolyte solution into which a protein binds in multilayers via multipoint binding or dissolves. The multilayer binding of a protein leads to high-capacity protein recovery and high-density enzyme immobilization [1]. When a hydrophobic moiety coexists with a positively charged moiety in the graft-chain phase, extractants that have both hydrophobic and negatively charged groups are impregnated or dissolve into the graft-chain phase. Impregnation of extractants enables a highly efficient complex to form with target metal ions in the absence of organic solvents [2].

A graft chain prepared by radiation-induced graft polymerization shows a nonuniform distribution of molecular masses. Characterization of a graft chain is difficult because the graft chain cannot be quantitatively isolated from the trunk polymer. However, a simple method to determine the mole percentage of the polymer brush was proposed for a poly (glycidyl methacrylate) chain grafted onto a polyethylene porous sheet [3]. Because a poly (glycidyl methacrylate) graft chain exhibits a shrunken conformation in water, which is a poor solvent for the reagents in epoxy-ring opening reactions, a charged group can be preferably introduced into the epoxy group of the polymer brush as opposed to the polymer root. This "part-selective" introduction of charged groups provides information on the boundary between the polymer brush and the polymer root. This method is applicable to partial introduction of functional groups into a graft chain, which confers the graft chain on a porous membrane the size-exclusion property.

2.1   Distribution of Functional Groups Along Graft Chains

2.1.1   Control of Order of Reagent Introduction [4]

Glycidyl methacrylate (GMA) was graft-polymerized onto an electron-beam-irradiated polyethylene porous hollow-fiber membrane. Subsequently, ion-exchange and chelate-forming groups, and hydrophobic and affinity ligands were introduced into the epoxy group of graft chain by reactions with various reagents. Thus far, the resultant membrane adsorbers have shown separation modes based on ion-exchange or electrostatic, chelating, hydrophobic, and affinity interactions, but not a size-exclusion mode. We have successfully prepared a porous hollow-fiber membrane capable of size exclusion or size recognition simply by designing the distribution of the functional groups along a graft chain.

The epoxy group of a poly-GMA chain grafted onto a porous hollow-fiber membrane was ring-opened with trimethylamine ($N(CH_3)_3$) and water to form trimethylammonium (TMA) and diol groups, respectively. Two functional groups were successively introduced into the epoxy groups. Two types of ion-exchange porous hollow-fiber membrane were prepared via two schemes, as shown in Fig. 2.1, to evaluate the binding capacities of phosphate ions and bovine serum albumin (BSA) in equilibrium with their respective feed concentrations, in accordance with the order of use of the two reagents, the resultant porous hollow-fiber membranes were referred to as TMA(x)-diol and diol-TMA(x) fibers, where x designates the molar conversion of the epoxy group to the TMA group. Here, the degree of GMA grafting was set at 170%, which corresponds to an epoxy group density of 11 (= 1700/142) mol/kg of the porous hollow-fiber membrane as a trunk polymer.

A solution containing phosphate ions or BSA was forced to permeate through the TMA(x)-diol or diol-TMA(x) fiber radially outward from the inside surface of the hollow fiber to the outside surface. Phosphate ions or BSA in the effluent penetrating the outside surface of the hollow fiber was continuously determined. From the breakthrough curve, i.e., effluent concentration versus effluent volume, the equilibrium binding capacity of the hollow fiber for phosphate ions or BSA was evaluated using Eqs. (2.1–2.4).

The TMA group serves as a strongly basic anion-exchange group to adsorb phosphate ions and BSA. A significant difference in the equilibrium binding capacity of BSA was observed between the TMA(x)-diol and diol-TMA(x) fibers (Fig. 2.2a). For BSA, which is a larger anion, the equilibrium binding capacity of the TMA(x)-diol fiber increased with increasing x, whereas the diol-TMA(x) fiber did not adsorb BSA until an x value of 70% was reached. In contrast, the equilibrium binding capacities of the two hollow fibers for phosphate ions, which are smaller anions, increased in proportion to x. The linear dependence of the equilibrium binding capacity of the TMA(x)-diol fiber for phosphate ions on x agreed well with that of the diol-TMA(x) fiber (Fig. 2.2b).

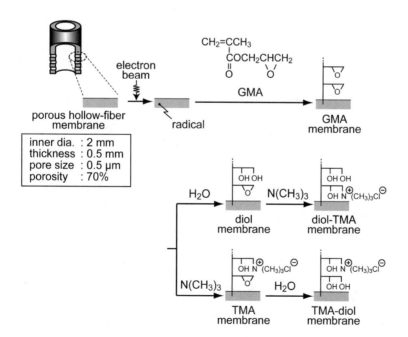

Fig. 2.1 Schemes for introduction of two functional groups into graft chain

The diol-TMA(x) fiber recognized the sizes of anions: BSA was excluded by the graft chain, whereas phosphate ions were included by the graft chain. This size-recognition capability can be explained by the distinct distribution of the two functional groups along the graft chain. The diol groups are exclusively introduced into the upper part of the graft chain resulting in an unswollen conformation, whereas the TMA groups introduced into the lower part cause the graft chain to expand owing to mutual electrostatic repulsion. This "two-layer structure of the graft chain," as illustrated in Fig. 2.3, leads to the size-based recognition of molecules and ions. Simply by designing the order of reagents introduced into the graft chain, a size-exclusive polymer brush was achievable. In principle, the difference in the degree of swelling of the graft chain in reagent solutions contributes to the differentiation of functional groups along the graft chain.

2.1.2 Selection of Solvent for Reagent for Functionalization [3, 5, 6]

A polyethylene porous sheet was adopted as a trunk polymer for grafting GMA. The permeability of the GMA-grafted porous sheet with a degree of GMA grafting of 220% was higher than that of the trunk porous sheet. Some part of the graft chain (the polymer root) invades the polyethylene matrix to swell it, and the remaining part of the graft chain (the polymer brush) locates onto the pore surface.

Water was employed as a solvent for the introduction of functional groups into the poly-GMA-grafted chain, although the addition of isopropyl alcohol to water was effective in accelerating the reaction: a mixture of sodium hydrogen sulfite (NaHSO$_3$) and sodium sulfite (Na$_2$SO$_3$) was dissolved in water as a solvent for the

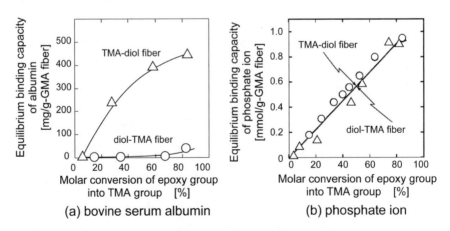

Fig. 2.2 Equilibrium binding capacities of TMA-diol and diol-TMA fibers for bovine serum albumin and phosphate ion. Reprinted with permission from Ref. [4]. Copyright 2009 The Membrane Society of Japan

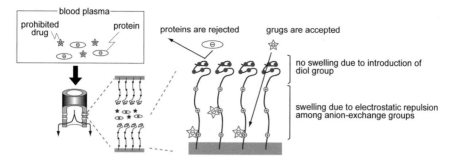

Fig. 2.3 Two-layer structure along the graft chain

introduction of a sulfonic acid group into the graft chain. Similarly, trimethylammonium chloride was dissolved in water as a solvent for the introduction of a trimethylammonium chloride group. Time courses of the molar conversion of the epoxy groups to sulfonic acid and trimethylammonium chloride groups are shown in Fig. 2.4. Through both modifications, the resultant ion-exchange porous sheets exhibited a peculiar swelling behavior: initially, the volume of the porous sheet remained unchanged, and beyond a specific value of molar conversion the porous sheet started to linearly swell (Fig. 2.5).

This behavior can be explained by the progression of the reaction interface between the graft chain and the reagent solution. The introduction of the sulfonic acid group at the reaction interface renews the reaction interface: the renewed reaction interface shifts beneath it along the graft chain. This shift continues until the ring opening of the poly-GMA chain with the reagents is completed, as illustrated in

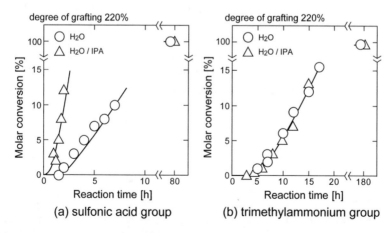

Fig. 2.4 Time courses of molar conversions for solvents for introduction of functional groups. Reprinted with permission from Ref. [3]. Copyright 2013 The Society of Chemical Engineers, Japan

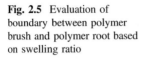

Fig. 2.5 Evaluation of boundary between polymer brush and polymer root based on swelling ratio

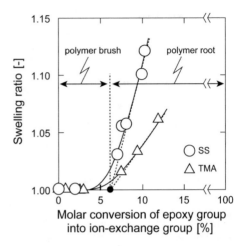

Fig. 2.6. The shifting reaction interface crosses the boundary between polymer brush and the polymer root.

During the reaction of the epoxy group of the polymer brush with NaHSO$_3$, the polymer brush extends from the pore surface owing to mutual electrostatic repulsion; the total volume of the porous sheet remains constant. As the sulfonic acid group starts to be introduced into the polymer root, the polyethylene chain surrounding the polymer root forces the porous sheet to swell. When the molar conversion reaches 100%, a specific value of molar conversion is equivalent to a boundary molar conversion in mole percentage, which means the mole percentage of polymer brush in the total graft chain.

The ratio of the polymer brush to the polymer root was 6–94 for the poly-GMA chain grafted onto a polyethylene porous sheet that was irradiated with an electron beam at a dose of 200 kGy; subsequently, GMA was graft-polymerized with 5 (v/v)% GMA/methanol solution, as shown in Fig. 2.5. A polymer root percentage of as high as 94% swelled the matrix of the porous sheet, resulting in the increases in pore size and flux. In contrast, a polymer brush with a percentage of as small as 6% held proteins in multilayers because the total length of the graft chain was estimated to be at least 1 μm. Metal ions were also captured by both the polymer

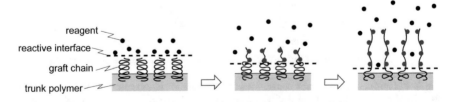

Fig. 2.6 Introduction of functional groups into GMA graft chain in aqueous solvent

brush and polymer root because metal ions do not differentiate the polymer brush from the polymer root.

When the GMA graft chain as a hydrophobic polymer chain is graft-polymerized onto polyethylene as a hydrophobic matrix, the reaction interface shifts downward from the upper front of the graft chain when water is used as a solvent of functionalization for the epoxy ring (Fig. 2.6). A marked change in the volume of the grafted material reflects the reaction interface crossing the boundary between the polymer brush and the polymer root.

2.1.3 Control of Functional Group Distribution by pH of Reagent Solution [7]

Distribution of phenyl groups as hydrophobic ligands along a graft chain varied as a function of the pH of the phenol solution. Phenyl groups were introduced into a poly-GMA chain grafted onto a porous hollow-fiber membrane with a degree of GMA grafting of 150% (Fig. 2.7). The pH of the phenol solution was in the range from 9 to 13. As illustrated in Fig. 2.8, at a low pH, the phenol solution is relatively hydrophobic because phenol does not dissociate, whereas at a high pH, the phenol solution is relatively hydrophilic because phenol dissolves as the phenoxide ion. The phenyl groups are introduced into the graft chain in swollen and unswollen states at low and high pH, respectively. In other words, uniform and partial distributions of phenyl groups are possible along the graft chain. The resultant phenyl-group-containing porous hollow-fiber membranes are referred to as hydrophobic porous hollow-fiber membranes.

When a protein dissolved in a buffer containing 2 M ammonium sulfate $((NH_4)_2SO_4)$ permeated through a hydrophobic porous hollow-fiber membrane, the protein was bound to the phenyl group of the graft chain on the basis of hydrophobic interactions. In contrast, the permeation of ammonium sulfate-free buffer through the membrane causes the elution of the protein adsorbed onto the membrane. The phenyl-group-containing porous hollow-fiber membrane prepared with a high-pH phenol solution exhibited a low elution percentage of protein, where the elution percentage was defined as the percentage of the amount of the eluted protein with

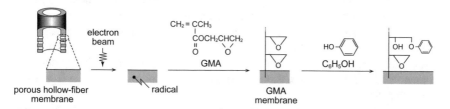

Fig. 2.7 Introduction of phenyl groups as hydrophobic ligands into porous hollow-fiber membrane

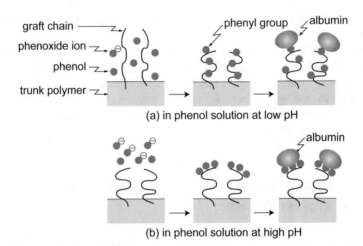

Fig. 2.8 Distribution of phenyl groups along graft chain at various pHs of phenol solution. Reprinted with permission from Ref. [7]. Copyright 1996 Elsevier

respect to that of the adsorbed protein. This dependence of elution percentage on the pH of the reagent solution used to introduce functional groups into the GMA graft chain reflects the distribution of the functional groups. The change in the dissolved form of the reagent with pH governs the swelling of the graft chain, resulting in the distribution of functional groups and ligands.

2.2 Detection of Extension and Shrinkage of Graft Chain

A graft chain was appended uniformly across a polyethylene-made porous hollow-fiber membrane with a thickness in the range of 0.6–0.7 mm,an average pore diameter of 0.4 μm and a porosity of 70%. An epoxy-group-containing vinyl monomer as a precursor monomer, GMA, was readily graft-polymerized to yield a graft chain mass of 1 kg/kg of the starting membrane. This means that approximately 7 mol of epoxy groups was added to the membrane. Subsequently, some of the epoxy groups were converted to ion-exchange or chelate-forming groups to obtain novel polymeric adsorbents with a feasible density of functional groups higher than commercially available ion-exchange or chelate-forming resin.

We were concerned that the pores were filled with grafted polymer chains with the same mass as the starting porous polyethylene hollow-fiber membrane. Fortunately, the porous structure of the membrane was retained after grafting. As shown in Fig. 2.9, the pores were enlarged, resulting in an increase in liquid permeability [8].

The graft chains without a crosslinked structure can shrink or extend because one end of the graft chain is free while the other end is tethered to the trunk

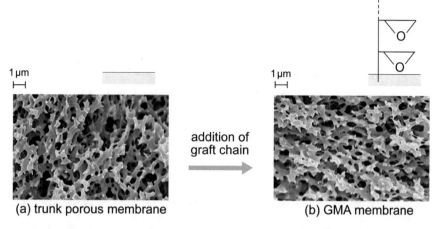

Fig. 2.9 SEM image of cross section of porous hollow-fiber membrane

polymer. The conformation of the graft chains in liquids is governed by internal and external factors: the internal factors include the length and density of the graft chain, and the functional group density introduced into the graft chain; the external factors include the pH, ionic strength, temperature, and viscosity of the liquid surrounding the graft chain. The graft chain responds to the external factors in a static and dynamic manner.

2.2.1 Ionic Crosslinking of Graft Chains [9]

First, GMA was graft-polymerized onto a porous hollow-fiber membrane with a degree of GMA grafting of 220%. The epoxy groups of the graft chain were converted to sulfonic acid groups with a molar conversion of 28% (Fig. 2.10a). The sulfonic acid group ($-SO_3H$) was appended uniformly throughout the membrane thickness. The end of the 10-cm-long hollow fiber was connected to a syringe containing a feed solution, and the other end was sealed. A buffer containing protein was forced to permeate radially outward from the inside surface of a hollow fiber.

The fluxes of three hollow fibers, i.e., SS-diol, diol, and PE fibers, for various liquids are listed in Table 2.1. The flux was evaluated by dividing the flow rate of liquid permeating through the pores by the inside surface area of the hollow fiber at a constant permeation pressure and liquid temperature. A high flux is favorable for high-throughput operation in industrial uses of porous hollow-fiber membranes. The flux of the SS-diol fiber for water was much lower than those of the diol and PE fibers. The SO_3H-group-containing polymer brush swells owing to mutual electrostatic repulsion among the $-SO_3H$ groups and narrows the pores.

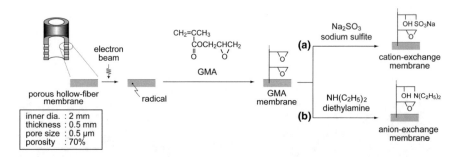

Fig. 2.10 Introduction of cation- and anion-exchange groups into porous hollow-fiber membranes

Table 2.1 Permeation flux of porous hollow-fiber membranes for various salt solutions (25 °C, 0.06 MPa)

Solution	SS-diol membrane	Diol membrane	PE membrane
Water	0.10	1.9	3.1
0.005 mol/L NaCl	0.19	2.0	3.1
0.005 mol/L KCl	0.16	2.0	3.1
0.005 mol/L MgCl$_2$	1.2	2.0	3.1
0.005 mol/L CaCl$_2$	1.5	2.0	3.1

The fluxes of the SS-diol fiber for 0.005 M magnesium and calcium chloride solutions were respectively 12- and 15-fold higher than that for pure water. This remarkable restoration of liquid permeability can be explained by the shrinkage of the polymer brush induced by ionic crosslinking of magnesium and calcium ions between neighboring $-SO_3H$ groups. In contrast, the permeation of 0.005 M sodium chloride solution resulted in poor restoration of liquid permeability. It is reasonable that divalent cations, not monovalent cations, are effective in crosslinking the SO_3H-group-containing polymer brush. Permeation of divalent ion-containing solutions opened the pores by ionically crosslinking the SO_3H-group-containing graft chains. After the ionic crosslinking with magnesium and calcium ions as simple cations, lysozyme as macromolecular cation was entangled by the polymer brush.

One g/L lysozyme (Mr 14,000, pI 10.7) was fed to the inside surface of the SS-diol fiber at a constant permeation pressure of 0.06 MPa to determine the protein concentration and flow rate of the effluent penetrating the outside surface of the hollow fiber. After the adsorption reached equilibrium, the fiber was washed with the buffer and eluted with 1 M sodium chloride-containing buffer. Protein concentration and flux are shown in Fig. 2.11 as a function of effluent volume. As the adsorption of lysozyme onto the SS-diol fiber proceeded, the flux gradually increased. This reflects the shrinkage of the polymer brush caused by the entanglement of the protein via multipoint binding.

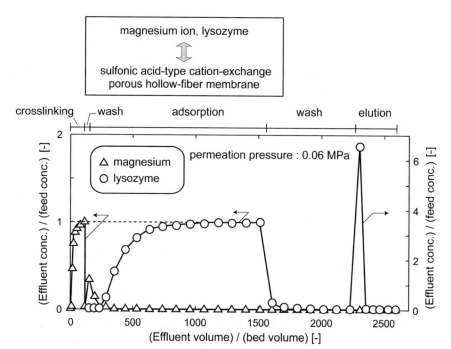

(a) breakthrough and elution curves of magnesium ion and lysozyme

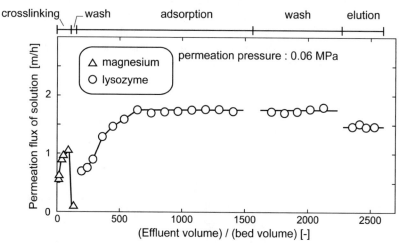

(b) variation in permeation flux for adsorption and elution of lysozyme

Fig. 2.11 Comparison of the amount of protein bound and the permeation flux. Reprinted with permission from Ref. [9]. Copyright 1999 Elsevier

When the effluent lysozyme concentration reached the feed lysozyme concentration, the equilibrium binding capacity for lysozyme was calculated using Eqs. (2.1–2.4) as 0.42 g/g of the SS-diol fiber. This value can be converted to a degree of multilayer binding of protein of 38 by dividing the equilibrium binding capacity by the theoretical monolayer binding capacity (0.011 g-lysozyme/g of fiber).

Multilayer binding of proteins is useful for protein recovery and enzyme immobilization because the graft chain provides a space with high capacity and activity. The pores of the SS-diol fiber are rimmed by the graft chains ionically crosslinked with magnesium ions; lysozyme as a macromolecular cation replaces magnesium ions as a simple cation to further crosslink the graft chain while diffusing into the depth of the polymer brush (Fig. 2.12). The graft-chain phase diffusion is driven by the gradient of the amount of adsorbed lysozyme.

Instead of sodium sulfate, taurine ($NH_2(CH_2)_2SO_3H$) was employed as the reagent for the epoxy-ring opening of the GMA graft chain [10]. Because taurine is an ampholyte having both sulfonic acid and amino groups, self-neutralization inhibited the extension of the polymer brush.

The effective length of the polymer brush can be estimated from the degree of multilayer binding of proteins. The term "effective" means the length of the polymer brush along which the proteins are entangled via multipoint binding with ion-exchange groups of the graft chain. The amount of proteins adsorbed was determined by permeating the protein solution through the pores rimmed by the ion-exchange polymer brush. Here, the term "ion exchange" means charged or ionizable. Proteins can be held by the polymer brush and cannot invade the polymer-root-embedded matrix because of their sizes; proteins are a probe for the presence of the polymer brush. On the other hand, simple metal ions cannot be a probe specific for the presence of the polymer brush.

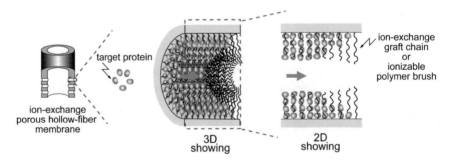

Fig. 2.12 Illustration of multilayer binding of protein to ion-exchange graft chain or ionizable polymer brush

2.2.2   Projection of Protein Adsorption and Elution to Liquid Permeability [11, 12]

Diethylamino groups ($-N(C_2H_5)_2$) were introduced into a polymer chain grafted onto a porous hollow-fiber membrane (Fig. 2.10b). The degree of grafting and the molar conversion of epoxy groups into diethylamino groups were 180 and 50%, respectively. A protein solution prepared from bovine blood was forced to permeate through the pores rimmed by the anion-exchange polymer brush radially outward from the inside surface of the hollow fiber to the outside surface.

A real protein solution instead of a protein solution model was used as a feed. Proteins in the feed analyzed by gel electrophoresis are shown in Fig. 2.13. A number of bands corresponding to various proteins were detected. Almost all the proteins were bound to the anion-exchange polymer brush before gelsolin as a target protein was specifically eluted with a specific eluent. This protein purification method is referred to as affinity elution.

Gelsolin has molecular mass of 90,000 and pI of 5.8. Gelsolin is contained in blood and decomposes actin produced following the degradation of muscle tissue. It plays an essential role in preventing the blocking of blood vessels. Of the proteins bound to the polymer brush, gelsolin was specifically eluted by permeating 2 mmol/L calcium chloride solution through the pores because gelsolin is a calcium-binding protein.

The highly selective binding of calcium ions to gelsolin induces changes in the conformation and surface charge of gelsolin, resulting in the stripping of gelsolin

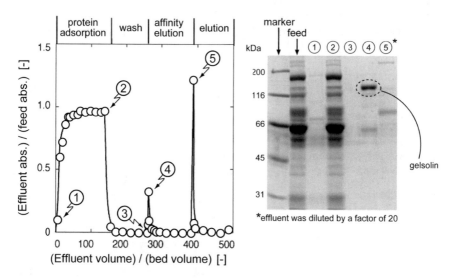

Fig. 2.13 Ion-exchange adsorption of gelsolin onto anion-exchange porous hollow-fiber membrane and affinity elution from hollow fiber with calcium ions. Reprinted with permission from Ref. [11]. Copyright 2005 Elsevier

from the anion-exchange polymer brush. The remaining proteins bound to the anion-exchange polymer brush were quantitatively eluted with 1 M sodium chloride.

Gelsolin purification consists of a series of six steps: (1) adsorption of total proteins from protein solution derived from bovine blood, (2) washing with a buffer, (3) affinity elution with calcium ion-containing buffer, (4) washing with the buffer, (5) elution with sodium chloride-containing buffer, and (6) washing with the buffer. With the execution of a series of procedures, the flux, total protein concentration, and protein profile of the effluent penetrating the outside surface of the hollow fiber vary, as shown in Fig. 2.13. First, the flux at a constant permeation pressure decreased as the protein adsorption proceeded. Second, the flux increased by 10% after the affinity elution of gelsolin. Finally, the flux returned to its initial value following the quantitative elution of the remaining proteins. Restoration of flux means that the amount of proteins nonselectively adsorbed onto the anion-exchange porous hollow-fiber membrane was negligibly small.

2.2.3 Expansion and Shrinkage of Graft Chain Accompanied by Protein Binding [13]

Sulfonic acid groups ($-SO_3H$) or diethylamino groups ($-N(C_2H_5)_2$) were introduced into the GMA-grafted porous hollow-fiber membrane with a degree of grafting of 160%. Representative protein solutions were forced to permeate through the pores of the porous ion-exchange hollow-fiber membrane, and changes in liquid permeability as protein binding progressed were evaluated. A marked difference was observed between two types of hollow fiber (Fig. 2.14): lysozyme binds to the

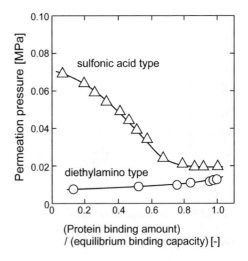

Fig. 2.14 Changes in permeation pressure accompanied by protein binding to sulfonic acid group- or diethylamino group-containing graft chains. Reprinted with permission from Ref. [13]. Copyright 2000 American Chemical Society

sulfonic acid-containing polymer brush to shrink it; in contrast, bovine serum albumin binds to the diethylamino group-containing polymer brush to expand it.

Following radiation-induced graft polymerization, the graft chain was appended uniformly across porous polymeric materials (e.g., a porous hollow-fiber membrane and a porous sheet made of polyethylene): a controlled amount of the graft chain can be confined in a porous structure. Because adsorption and elution of proteins induce the expansion and shrinkage of the graft chain, respectively, pore size is variable enough to control liquid permeability. When the graft chain is confined into a porous structure, liquid permeability becomes an index of the adsorption and elution of proteins.

Pore-size changes caused by conformational changes of the graft chain can be easily detected by determining the permeation pressure required for a liquid to permeate through a porous hollow-fiber membrane at a constant rate. Thus, we translate the liquid permeability changes of a modified porous hollow-fiber membrane to conformational changes of a graft chain.

2.3   Liquidlike Graft-Chain Phase

2.3.1   Liquidlike Behavior of Graft Chain as Catalytic Field [14, 15]

Sodium styrene sulfonate (SSS) was cografted with 2-hydroxyethyl methacrylate (HEMA) onto porous hollow-fiber polyethylene. Here, HEMA enhances the grafting of a highly hydrophilic monomer, SSS, onto a relatively hydrophobic trunk polymer, polyethylene [16]. When the functional group of the resultant graft chain was converted from a Na-form to a H-form, the resultant strongly acidic cation-exchange or SO$_3$H-group-containing graft chain worked as an acid catalyst. The -SO$_3$H group density of the graft chain was determined to be 2 mol/kg of the trunk polymer. This density was comparable to a concentration of 1 M sulfuric acid (H$_2$SO$_4$).

The catalytic activity of the SO$_3$H-group-containing graft chain was evaluated using the hydrolysis reaction of methyl acetate or sucrose. Methyl acetate (MW, 74) and sucrose (MW, 342) were reactant models for probing the accessibility of the substrate to the graft phase, that is, the density of the graft phase. For comparison, a commercially available strongly acidic cation-exchange bead (Amberlite IR-120B manufactured by Dow Chemical Co.) based on a styrene–divinylbenzene copolymer was selected. As illustrated in Fig. 2.15, the crosslinked structure of this copolymer contrasts with the non-crosslinked polymeric structure of the graft-chain phase.

First, the hydrolysis rate of methyl acetate was analyzed on the basis of the first-order reversible reaction.

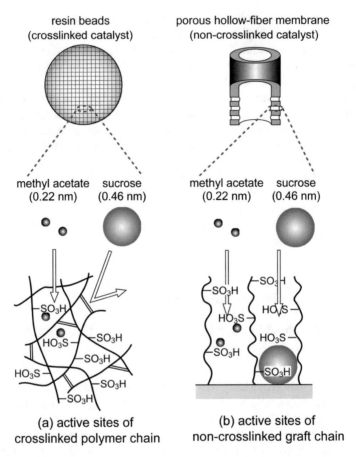

Fig. 2.15 Comparison of structures of polymeric acidic catalysts. Reprinted with permission from Ref. [14]. Copyright 1994 American Chemical Society

$$CH_3COOCH_3 + H_2O = CH_3COOH + CH_3OH \qquad (2.1)$$

The hydrolysis activities of two SO_3H-group-containing polymeric catalysts were compared in a temperature range from 40 to 55 °C: (1) one catalytic site is the $-SO_3H$ group of the polymer chain grafted onto a porous polyethylene hollow-fiber membrane and (2) the other catalytic site is the $-SO_3H$ group introduced into the styrene–divinylbenzene copolymer. The rate constants determined at different reaction temperatures were plotted according to the following Arrhenius equation:

$$k = A\exp[-\Delta E/(RT)], \qquad (2.2)$$

where k is the rate constant of the reaction, and A and ΔE are the frequency factor and activation energy, respectively. R and T are the gas constant and reaction temperature, respectively.

As shown in Fig. 2.16a. The Arrhenius plots overlapped between two $-SO_3H$ groups of the non-crosslinked polymer chain and crosslinked polymer chain. Furthermore, these values obtained from a heterogeneous system are consistent with the reaction rate constant for sulfuric acid (H_2SO_4) as a homogeneous system. Methyl acetate as a substrate does not distinguish the $-SO_3H$ group of the graft chain from the $-SO_3H$ group in the liquid. Namely, the graft-chain phase working as an active site of methyl acetate hydrolysis behaves like a liquid.

The hydrolysis rate of sucrose was analyzed as an irreversible first-order reaction.

$$\text{sucrose} + H_2O = \text{fructose} + \text{glucose} \tag{2.3}$$

The logarithm of the reaction rate constant is plotted against the reciprocal of reaction temperature ranging from 55 to 70 °C.

A straight line for the graft chain catalyst did not agree with that for nongraft chain catalyst (Fig. 2.16b). The k of the graft chain was higher than that of the nongraft chain. The slope of the straight line for the graft chain was steeper than that for the nongraft chain. For sucrose hydrolysis, the activation energies of the graft and nongraft chains were calculated to be 95 and 59 kJ/mol, respectively, using Eq. (2.2).

The magnitude of activation energy reflects the contribution of chemical and physical processes to the overall reaction rate. A low activation energy of the non-graft or crosslinked polymeric catalyst indicates that the overall reaction is governed by physical processes: a sucrose molecule diffuses in the phase consisting of the crosslinked polymer network and approaches the $-SO_3H$ group to be hydrolyzed into

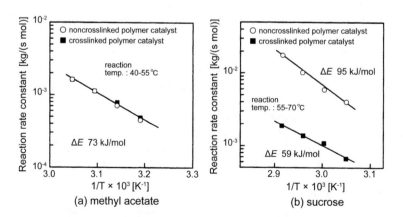

Fig. 2.16 Comparison of activation energy for hydrolysis between crosslinked and non-crosslinked polymer catalysts. Reprinted with permission from Ref. [14]. Copyright 1994 American Chemical Society

fructose and glucose. The higher activation energy of the graft-type polymeric catalyst means that intrinsic hydrolysis of sucrose is a rate-determining step. The
$-SO_3H$ group-containing graft-chain phase provides a space for a sucrose molecule
to easily diffuse: the sucrose molecule does not sense any mass-transfer resistance.
We therefore consider that the graft-chain phase behaves like a liquid.

The hydrolytic activity of the acid was compared between the $-SO_3H$ group
introduced into the flexible graft chain and that introduced into the rigid crosslinked
polymer network. Because the SO_3H-group-containing polymer chain is soluble in
water as it swells, the crosslinked polymer network is used as an insoluble support.
On the other hand, the SO_3H-group-containing polymer chain grafted onto an
insoluble trunk polymer extends itself in water to work as sulfuric acid does.

2.3.2 Liquidlike Behavior of Graft Chain for Extractant Impregnation [17, 18]

A graft chain containing hydrophobic ligands such as the octadecyl group
($-C_{18}H_{37}$) is useful for impregnating extractants specific for target metal ions.
Extractants are categorized into three groups: acidic, neutral, and basic. Most
extractants contain a hydrophobic domain; therefore, a hydrophobic graft chain will
interact with the extractants. The extractants are not immobilized on the graft chain
via covalent bonding but are impregnated onto the graft chain via hydrophobic
interactions. Therefore, the extractants can be mobile on the graft chain.

A scientific goal was achieved from a promising application of an
extractant-impregnated fiber to the recovery and purification of neodymium
(Nd) and dysprosium (Dy) ions. First, GMA was graft-polymerized onto an
electron-beam-irradiated nylon-6 fiber. Second, some of the epoxy groups of the
resultant graft chain were converted to dodecylamino groups ($-NHC_{12}H_{25}$) as
hydrophobic ligands. Third, HDEHP (bis(2–ethylhexyl) phosphate) was impregnated onto the hydrophobic graft chain. The resultant fiber is referred to as an
HDEHP-impregnated fiber. As shown in Fig. 2.17, the HDEHP molecule is not
linked to the graft chain via covalent binding but is held via hydrophobic interactions between dodecylamino groups of the graft chain and 2–ethylhexyl groups in
HDEHP.

The amount of impregnated HDEHP was 0.43 mmol/g of the product fiber. The
HDEHP-impregnated fiber was immersed in 0.01 to 0.1 M hydrochloric acid
containing 290 mg-Nd/L and 30 mg-Dy/L until adsorption equilibrium was
attained. The distribution coefficient D was defined as follows:

$$D = (\text{amount of Nd or Dy adsorbed per mass of fiber})$$
$$/ (\text{concentration of Nd or Dy in aqueous solution}) \tag{2.4}$$

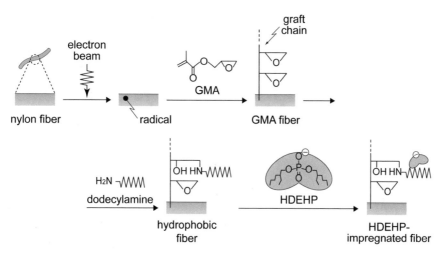

Fig. 2.17 Scheme of HDEHP impregnation into a polymer chain grafted onto nylon fiber

For comparison, HDEHP was dissolved in n-dodecane to yield a 0.5 M HDEHP-containing oil phase. This oil phase was placed in contact with a Nd or Dy aqueous phase until extraction equilibrium was attained. In the calculation of the distribution coefficient, the amount of Nd or Dy extracted per volume of the oil phase was used in Eq. (2.4) instead of the amount of Nd or Dy adsorbed per mass of fiber.

Distribution coefficients are shown in Fig. 2.18 as a function of hydrochloric acid concentration. A straight line with a slope of -3 was obtained for the indicated range of hydrochloric acid concentrations. The slope can be explained by the following reversible reaction:

$$3(HA)_2 + M^{3+} = ((HA)_2)_3 M + 3H^+, \qquad (2.5)$$

where HA and $(HA)_2$ denote HDEHP and its dimer, respectively. Then, an equilibrium constant K is defined as

$$K = \left(\left[((HA)_2)_3 M \right] [H^+]^3 \right) / \left([(HA)_2]^3 [M^{3+}] \right). \qquad (2.6)$$

where the components in brackets represent the respective concentrations. From this equation, the distribution coefficient D is derived as

$$\log D = \log K + 3 \log \left[(HA)_2 \right] - 3 \log[H^+] \qquad (2.7)$$

On the other hand, an oil phase consisting of HDEHP and n-dodecane was equilibrated with an aqueous phase to acquire the distribution coefficient. As shown in Fig. 2.18, results of solid–liquid extraction agreed well with those of liquid–liquid

Fig. 2.18 Dependence of
distribution coefficient of
HDEHP-impregnated fiber for
neodymium and dysprosium
on HCl concentration

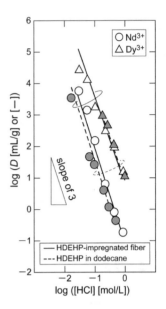

extraction. For example, the Ds of dysprosium of HDEHP-impregnated fibers and
HDEHP dissolved in n-dodecane were $10^{2.03}$ and $10^{2.05}$, respectively.

Impregnated HDEHP with dodecylamino groups as hydrophobic ligands on a
graft chain exhibited a similar characteristic in metal extractions to HDEHP dissolved
in n-dodecane as an oil phase: for metal ions in aqueous solution, the graft-chain
phase behaves like the oil phase, as illustrated in Fig. 2.19. The graft-chain phase
consisting of hydrophobic graft chains provides an interface similarly to an organic
solvent. HDEHP is not immobilized on the graft chain via covalent binding but
moves along the hydrophobic ligands of the graft chain to capture metal ions.

The transport of extractants along the hydrophobic graft chain driven by the
gradient of the amount of impregnated extractant is similar to the transport of
proteins along an ion-exchange graft chain driven by the gradient of the amount of

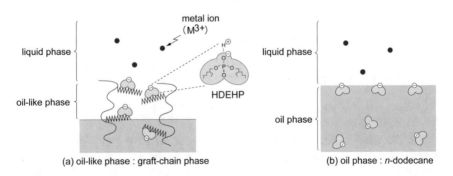

Fig. 2.19 Graft-chain phase extraction versus liquid–liquid extraction

adsorbed proteins. The graft-chain phase diffusion allows molecules and ions to be transported at much higher rates than ordinary diffusion in a liquid because their transport is enhanced by the functional groups of the graft chain.

The amount of HDEHP impregnated onto the nylon fiber was 0.43 mmol/g of the product fiber, which was 48% of that of commercially available HDEHP-impregnated beads (LEWATIT VP OC1026). However, high performance in elution chromatography is achievable because of short diffusional mass-transfer paths along the HDEHP-impregnated fiber compared with HDEHP-impregnated beads. The HDEHP-impregnated fiber was packed into a column with an inner diameter of 13 mm and a packed height of 11 mm. Two rare-earth metal (REM) ions, Nd and Dy, were loaded on the column at the top end. The Nd and Dy compositions of the solutions were equivalent to those of a cutting powder solution from a neodymium magnet.

Elution chromatograms were acquired in a stepwise mode. Hydrochloric acid as an eluent at three concentrations, i.e., 0.2, 0.3, and 1.5 M, flowed through the column. Nd and Dy ions were quantitatively eluted using 0.2 and 0.3 M hydrochloric acid, respectively. For comparison, LEWATIT beads were packed into the same column at the same packing height as the HDEHP-impregnated fiber. Elution chromatograms of the two columns are shown in Fig. 2.20.

The widths of the first and second peaks corresponding to Nd and Dy ions, respectively, of the HDEHP-impregnated fiber-packed column were smaller than those of the HDEHP-impregnated bead-packed column. The fiber-packed column is advantageous over the bead-packed column in that the higher peak and the narrower tail indicate higher concentrations of Nd and Dy ions in the eluent.

2.3.3   Diffusion in Graft-Chain Phase [19]

Sucrose recognizes a sulfonic acid-containing graft-chain phase as sulfuric acid, whereas HDEHP recognizes a dodecylamino graft-chain phase as n-dodecane. These graft-chain phases behave like liquids. Diffusion of molecules or ions adsorbed on the graft chain or in the graft-chain phase is driven by the gradient of the amount of adsorbed molecules or ions just as their diffusion in a liquid is based on the concentration gradient. We referred to this diffusion as graft-phase diffusion. This idea may be analogous to surface diffusion proposed for the high-speed diffusion of organic compounds adsorbed onto activated carbon.

To demonstrate the graft-chain diffusion, copper ions adsorbed onto an iminodiacetate-type chelating porous hollow-fiber membrane were adopted. Using the experimental apparatus in the permeation mode (Fig. 1.14), copper ions were bound until the chelating group corresponding to one-half the membrane thickness was saturated. Subsequently, the hollow fiber was immersed in water. The profile of copper ions adsorbed across the membrane thickness was determined as a function of immersion time. No copper ions were detected in the water during the immersion

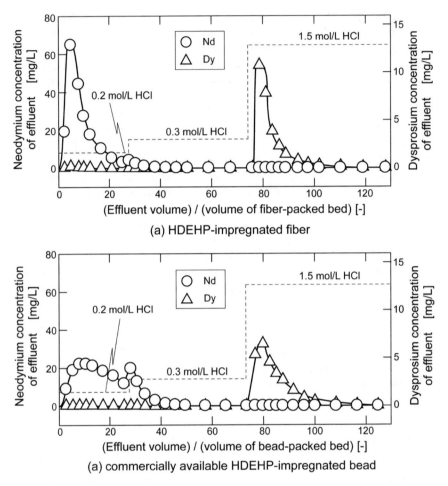

Fig. 2.20 Comparison of elution chromatograms between HDEHP-impregnated fiber- and bead-packed beds

of the hollow fiber. That is, the total amount of copper ions adsorbed onto the hollow fiber remained constant.

The time courses of the profiles of adsorbed copper ions are shown in Fig. 2.21. The transport rate of adsorbed copper ions cannot be explained by ordinary diffusion in the liquid in pores: the observed transport rate of copper ions was much higher than the transport rate estimated from the ordinary diffusion of copper ions in the liquid in pores. Thus, we postulated that graft-chain diffusion is driven by the gradient of the amount of adsorbed copper ions along the graft chain (Fig. 2.22). The graft-chain diffusive flux J_g is expressed by

Fig. 2.21 Graft-chain phase diffusion of metal ions in the direction of hollow-fiber membrane thickness

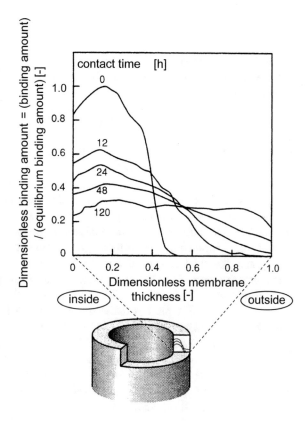

Fig. 2.22 Graft-chain phase diffusion

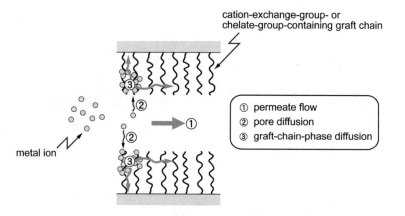

$$J_g = -D_g \, \mathrm{d}q/\mathrm{d}z, \tag{2.8}$$

where D_g is the diffusivity of the graft-chain phase. The term $\mathrm{d}q/\mathrm{d}z$ is the gradient of the amount of adsorbed copper ions. The ordinary diffusive flux J can be expressed by Fick's law as follows:

$$J = -D\mathrm{d}C/\mathrm{d}z, \tag{2.9}$$

where D is the diffusivity in a liquid, and C is the concentration.

When the graft-chain phase behaves like a liquid, crosslinking of the graft chain results in an increase in the viscosity of the phase. At a high viscosity, the diffusivity of molecules and ions in the graft-chain phase is low.

GMA was appended uniformly onto a porous sheet with a thickness of 1 mm, followed by the introduction of iminodiacetate groups into the epoxy groups. The resultant pore rimmed by the chelating graft chain had a diameter of approximately 1 μm, and the thickness of the graft-chain phase was approximately 0.1 μm. Therefore, the ratio of the length to the thickness of the graft-chain phase for mass transfer of copper ions to the chelating group was 10^3 to 0.1. By crosslinking the chelating graft chain with a crosslinker, the shape of the breakthrough curve was improved, as shown in Fig. 2.23. Only by adding the crosslinker to GMA at a concentration of 1% in graft polymerization, the practical or breakthrough binding capacity of the sheet for copper ions was increased by 20%. The crosslinking of the graft chain inhibits the axial mixing of copper ions along the pore length or sheet thickness. The graft-phase diffusion contributes to the increase in mass-transfer rate or decrease in separation efficiency.

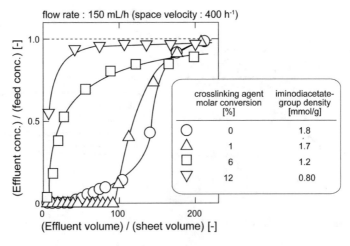

Fig. 2.23 Effect of crosslinking of graft chains containing iminodiacetate groups on breakthrough curve

2.4 Radical Storage for Preirradiation Grafting [20]

We have adopted polyethylene as the trunk polymer for radiation-induced graft polymerization. Polyethylene is advantageous in that it is physically stable against irradiation and economically feasible for commercialization, and polyethylene can be formed into porous hollow-fiber membranes, porous sheets, films, and particles. Polyethylene is classified into low-density (LD), high-density (HD), and ultrahigh molecular weight (UHMW) polyethylene according to its density.

Polyethylene consists of crystalline and amorphous domains. A polyethylene chain may be folded to form two-dimensional crystals called lamellae. Lamellae merge to form crystalline domains. On the other hand, unfolded polyethylene forms amorphous domains. Transitional domains exist between the crystalline and amorphous domains.

Irradiation of polyethylene with an electron beam or gamma rays collapses some of the carbon-hydrogen bonds, resulting in the generation of radicals. Alkyl, allyl, and peroxy radicals have been observed by electron spin resonance (ESR) spectroscopy. Among these radicals, alkyl radicals were found to contribute to the initiation of graft polymerization.

The energy of an electron beam generated at appropriate electrical voltage and current governs the depth of irradiation into polyethylene. Our goal is to prepare polymeric adsorbents with high capacity and rate of binding; therefore, we selected a dose so that the radicals can be produced uniformly throughout a polyethylene matrix by high-energy beam irradiation. The beam cannot distinguish crystalline domains from amorphous domains.

During the contact with a previously degassed vinyl monomer solution, the vinyl monomer invades the matrix to initiate graft polymerization from the radicals produced, as illustrated in Fig. 2.24. The vinyl monomer cannot invade the interstices among the folded polyethylene chains of the lamellae because it has a larger molecular size than the interstices. When the irradiated polymer comes in contact with air, oxygen diffuses into the matrix to react with the radicals in the amorphous domains and the radicals in the crystalline domains. The reaction of oxygen with the radicals forms peroxy radicals that are less active than alkyl radicals.

Sodium styrene sulfonate (SSS) was graft-polymerized onto a high-density polyethylene film to prepare a cation-exchange membrane for electrical dialysis used in manufacturing salt from seawater. The polymeric structure of the high-density polyethylene film, including factors such as crystallite size, was characterized by small-angle neutron scattering (SANS). We have thus far understood that the irradiation of crystalline domains only provides radicals on the crystallite surface without changing the crystallite structure. However, the size of the crystallites was reduced by the graft chain, whereas their size remained constant owing to the subsequent chemical modification [21]. Some of the folded polyethylene chains collapse as a result of graft polymerization.

Fig. 2.24 Radical behavior in the crystalline and amorphous domains of polyethylene

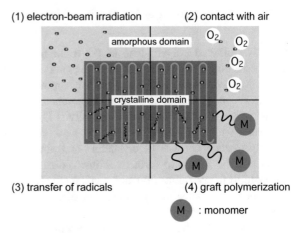

(1) electron-beam irradiation   (2) contact with air

amorphous domain

crystalline domain

(3) transfer of radicals   (4) graft polymerization

M : monomer

2.5 Stable Immobilization of Precipitates by Graft Chains [22, 23]

Vinyl benzyl trimethylammonium chloride (VBTAC) and *N*-vinyl-2-pyrrolidone (NVP) were cografted onto an electron-beam-irradiated nylon fiber to prepare an anion-exchange fiber (Fig. 2.25). VBTAC has trimethylammonium groups as strongly basic anion-exchange groups, whereas NVP accelerates the grafting of VBTAC. As shown by SEM-EDX, chloride ions adsorbed onto the anion-exchange fiber were uniformly distributed in the entire cross section of the VBTAC/NVP-cografted fiber with a diameter of 35 μm; therefore, uniform grafting of VBTAC over the entire volume of the fiber was demonstrated.

Ferrocyanide ions, $Fe(CN)_6^{4-}$, were bound to the anion-exchange fiber by immersing the fiber in an aqueous solution of potassium ferrocyanide ($K_4[Fe(CN)_6]$) before the fiber was immersed in an aqueous solution of cobalt chloride ($CoCl_2$). The color of the fiber changed from light yellow to dark green, corresponding to the color characteristic of insoluble cobalt ferrocyanide impregnated into the fiber via precipitation. Cobalt ferrocyanide has an extremely low solubility product and is a sparingly soluble salt; generally, the term "insoluble" is added to the front of "cobalt ferrocyanide."

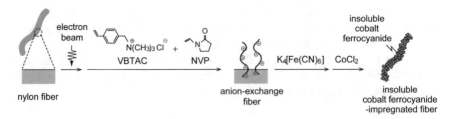

Fig. 2.25 Preparation scheme for insoluble cobalt-ferrocyanide-impregnated fiber

Ferrocyanide ions in solution in equilibrium with ferrocyanide ions bound to the anion-exchange group react with cobalt ions to form precipitates because of an extremely low solubility product. Subsequently, chloride ions bind to the anion-exchange groups in place of ferrocyanide ions. This successive precipitation continues until the adsorbed ferrocyanide ions are completely consumed. Before the precipitation, ferrocyanide ions were adsorbed uniformly across a cross section of the fiber. In contrast, the precipitate was immobilized at the periphery of the fiber, as shown in Fig. 2.26. SEM observation (Fig. 2.27b) shows that microparticles of cobalt ferrocyanide with an average size of 2 μm were scattered on the fiber surface.

No leakage of microparticles induced by the changes in pH and NaCl concentration was detected. This characteristic is favorable for the practical use of cobalt-ferrocyanide-impregnated fibers. The reason for a stable impregnation of insoluble cobalt ferrocyanide into these fibers was clarified from the results of rebinding of ferrocyanide ions to the cobalt-ferrocyanide-impregnated fibers [23]. The amount of ferrocyanide ions that rebound to the anion-exchange groups of the fibers decreased with increasing amount of cobalt ferrocyanide in the cobalt-ferrocyanide-impregnated fibers. This finding means that the precipitate deprives the anion-exchange groups of binding sites on the graft chain.

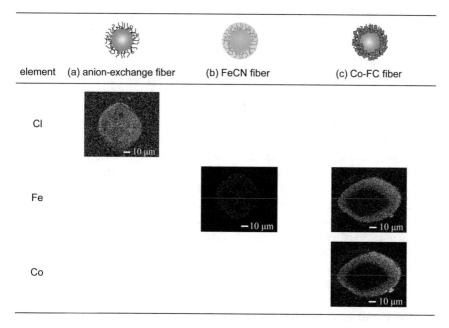

Fig. 2.26   Distribution of Cl, Fe, and Co across the diameters of different fibers. Reprinted with permission from Ref. [22]. Copyright 2015 Japan Radioisotope Association

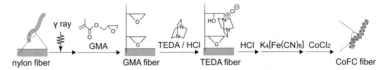

(a) impregnation of insoluble cobalt ferrocyanide with triethylenediamine (TEDA)-immobilized anion-exchange fiber

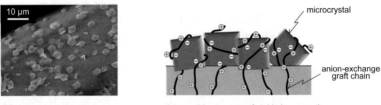

(b) SEM image of impregnated microcrystal   (c) possible structure of stable impregnation

Fig. 2.27 Electrostatic interaction between graft chains and insoluble cobalt ferrocyanide impregnated. Reprinted with permission from Ref. [23]. Copyright 2016 Taylor & Francis Ltd.

A possible impregnation structure of the microparticles on the graft chain is illustrated in Fig. 2.27c: microparticles with negative surface charges are entangled at multipoints and partially penetrated the graft chain with positively charged moieties. This entanglement prevents the leakage of microparticles into the external liquid in response to the pH or ionic strength changes as well as mechanical friction.

While an anion-exchange graft chain provides anionic species, e.g., $Fe(CN)_6^{4-}$, as one component of an inorganic precipitate, it binds the inorganic precipitate produced by reaction with a cationic species in the liquid, e.g., Co^{2+}, forming a stable impregnated structure (Fig. 2.28). The cobalt-ferrocyanide-impregnated fiber has already been applied to the removal of radioactive cesium ions from contaminated water at TEPCO's Fukushima Daiichi Nuclear Power Plant. For example, a braid of the fiber was utilized in a side ditch.

Charged polymer chains grafted to a nylon fiber can extend in aqueous solutions because of mutual electrostatic repulsion of the charged groups. The degree of extension depends on the type of charged group and the pH and ionic strength of the aqueous solution. This extension provides space for both precipitation and immobilization of inorganic compounds. This principle of impregnation is applicable to other impregnation systems. For example, sodium titanate precipitates were impregnated into nylon fibers for the removal of radioactive strontium ions from contaminated water.

Contaminated water from TEPCO's Fukushima Daiichi Nuclear Power Plant includes wastewater mainly stored in 1000-ton tanks and seawater accumulating in seawater-intake areas. The fibers manufactured in the form of a bobbin are woven

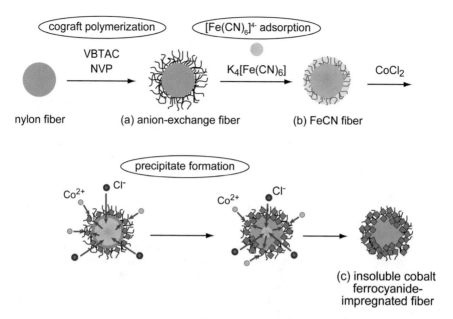

Fig. 2.28 Impregnation of insoluble cobalt ferrocyanide with anion-exchange fiber. Reprinted with permission from Ref. [22]. Copyright 2015 Japan Radioisotope Association

into a braid or a wound filter suitable for efficient and safe contact with contaminated water. Cobalt-ferrocyanide-impregnated fibers have an additional advantage over conventional adsorbents in that fibers bound to radioactive substances may be burned to reduce its volume before long-term storage.

References

1. T. Kawai, K. Saito, W. Lee, Protein binding to polymer brush, based on ion-exchange, hydrophobic, and affinity interactions. J. Chromatogr. B **790**, 131–142 (2003)
2. S. Asai, K. Miyoshi, K. Saito, Modification of a porous sheet (MAPS) for the high-performance solid-phase extraction of trace and ultratrace elements by radiation-induced graft polymerization. Anal. Sci. **26**, 649–658 (2010)
3. S. Uchiyama, R. Ishihara, D. Umeno, K. Saito, S. Yamada, H. Hirota, S. Asai, Determination of mole percentages of brush and root of polymer chain grafted onto porous sheet. J. Chem. Eng. Japan **46**, 414–419 (2013)
4. R. Shibahara, K. Hagiwara, D. Umeno, K. Saito, T. Sugo, Preparation of size-exclusion polymer chain grafted onto the pore surface of a porous hollow-fiber membrane. Membrane (Maku) **34**, 220–226 (2009)
5. R. Ishihara, S. Uchiyama, H. Ikezawa, S. Yamada, H. Hirota, D. Umeno, K. Saito, Effect of dose on mole percentages of polymer brush and root grafted onto porous polyethylene sheet by radiation-induced graft polymerization. Ind. Eng. Chem. Res. **52**, 12582–12586 (2013)

6. A. Nide, S. Kawai-Noma, D. Umeno, K. Saito, Reduction of buffer volume used in regeneration of anion-exchange porous hollow-fiber membrane by site-controlled introduction of anion-exchange group into graft chain. Membrane (Maku) **39**, 258–263 (2014)
7. N. Kubota, M. Kounosu, K. Saito, K. Sugita, K. Watanabe, T. Sugo, Control of phenyl-group site introduced on the graft chain for hydrophobic interaction chromatography. React. Polym. **29**, 115–122 (1996)
8. S. Matoba, S. Tsuneda, K. Saito, T. Sugo, Highly efficient enzyme recovery using a porous membrane with immobilized tentacle polymer chains. Nat. Biotechnol. **13**, 795–797 (1995)
9. N. Sasagawa, K. Saito, K. Sugita, S. Kunori, T. Sugo, Ionic crosslinking of SO₃H-group-containing graft chains helps to capture lysozyme in a permeation mode. J. Chromatogr. A **848**, 161–168 (1999)
10. K. Miyoshi, K. Saito, T. Shiraishi, T. Sugo, Introduction of taurine into polymer brush grafted onto porous hollow-fiber membrane. J. Membr. Sci. **264**, 97–103 (2005)
11. K. Hagiwara, S. Yonedu, K. Saito, T. Shiraishi, T. Sugo, T. Tojyo, E. Katayama, High-performance purification of gelsolin from plasma using anion-exchange porous hollow-fiber membrane. J. Chromatogr. B **821**, 153–158 (2005)
12. S. Yonedu, K. Saito, E. Katayama, T. Tojyo, T. Shiraishi, T. Sugo, Affinity elution of gelsolin adsorbed onto an anion-exchange porous membrane. Membrane (Maku) **30**, 269–274 (2005)
13. T. Kawai, K. Sugita, K. Saito, T. Sugo, Extension and shrinkage of polymer brush grafted onto porous membrane induced by protein binding. Macromolecules **33**, 1306–1309 (2000)
14. T. Mizota, S. Tsuneda, K. Saito, T. Sugo, Hydrolysis of methyl acetate and sucrose in SO₃H-group-containing grafted polymer chain prepared by radiation-induced graft polymerization. Ind. Eng. Chem. Res. **33**, 2215–2219 (1994)
15. T. Mizota, S. Tsuneda, K. Saito, T. Sugo, Sulfonic acid catalysts prepared by radiation-induced graft polymerization. J. Catalysis **149**, 243–245 (1994)
16. S. Tsuneda, K. Saito, S. Furusaki, T. Sugo, K. Makuuchi, Simple introduction of sulfonic acid group onto polyethylene by radiation-induced cografting of sodium styrenesulfonate with hydrophilic monomers. Ind. Eng. Chem. Res. **32**, 1464–1470 (1993)
17. T. Sasaki, S. Uchiyama, K. Fujiwara, T. Sugo, D. Umeno, K. Saito, Similarity of rare earth extraction by acidic extractantbis(2-ethylhexyl) phosphate (HDEHP) supported on a dodecylamino-group-containing graft chain and by HDEHP dissolved in dodecane. Kagaku Kogaku Ronbunsyu **40**, 404–409 (2014)
18. T. Sasaki, S. Uchiyama, K. Fujiwara, T. Sugo, D. Umeno, K. Saito, Nd/Dy resolution by SPE-based elution chromatography with bis(2–ethylhexyl) phosphate (HDEHP)-impregnated fiber-packed bed. Kagaku Kogaku Ronbunsyu **41**, 220–227 (2015)
19. G. Wada, R. Ishihara, K. Miyoshi, D. Umeno, K. Saito, S. Asai, S. Yamada, H. Hirota, Effect of chelating group density of crosslinked graft chain on dynamic binding capacity for metal ions. J. Ion Exchange **22**, 47–52 (2011)
20. K. Uezu, K. Saito, S. Furusaki, T. Sugo, I. Ishigaki, Radicals contributing to preirradiation graft polymerization onto porous polyethylene. Radiat. Phy. Chem. **40**, 31–36 (1992)
21. T. Miyazawa, Y. Asari, K. Miyoshi, D. Umeno, K. Saito, T. Nagatani, N. Yoshikawa, R. Motokawa, S. Koizumi, Development of novel ion-exchange membranes for electrodialysis of seawater by electron-beam-induced graft polymerization (IV) Polymeric structures of cation-exchange membranes based on nylon-6 film. Bull. Sea Water Sci. Jpn. **64**, 360–365 (2010)
22. M. Sugiyama, S. Goto, T. Kojima, K. Fujiwara, T. Sugo, D. Umeno, K. Saito, Impregnation process of insoluble cobalt ferrocyanide onto anion-exchange fiber prepared by radiation-induced graft polymerization. Radioisotopes **64**, 219–228 (2015)
23. S. Goto, S. Umino, W. Amakai, K. Fujiwara, T. Sugo, T. Kojima, S. Noma-Kawai, D. Umeno, K. Saito, Impregnation structure of cobalt ferrocyanidemicroparticles by the polymer chain grafted onto nylon fiber. J. Nucl. Sci. Tech. **53**, 1251–1255 (2016)

Chapter 3
Revolution in the Form of Polymeric Adsorbents 1: Porous Hollow-Fiber Membranes and Porous Sheets

Abstract Porous hollow-fiber membranes and porous sheets used for microfiltration can be modified into porous adsorbents by radiation-induced graft polymerization. The three-dimensional modification or modification over the entire volume of the porous trunk polymer provides a functional density comparable to that of conventional adsorbents. The ideal adsorption in a flow-through mode is achievable because the time required for a target to diffuse to the functional moiety is much shorter than the residence time of the target solution as it passes through the porous membrane or sheet. The multilayer binding of proteins via multipoints in the polymer brush is applied to the immobilization of an enzyme at a high density, leading to high activity in enzyme reactions such as the quantitative hydrolysis of 4 M urea solution.

Keywords Porous hollow-fiber membrane · Porous sheet · Three-dimensional modification · Ideal adsorption · Enzyme immobilization

3.1 Design Strategy for Functional Materials for Separation

3.1.1 Two Modes of Mass Transfer

Mass transport modes in liquids are divided into two categories: convective and diffusive. Convective flow can transport ions or molecules in forced liquid flow, whereas diffusive flow of ions or molecules is driven by concentration gradients. The term "diffusion" is widely used to refer to the result of mass dispersion by convection and diffusion; however, in this book, we strictly distinguish diffusion from convection.

Adsorbents are used to recover and remove ions and molecules from various liquids. Adsorbents in bead form with a diameter of approximately 0.5 mm have been prepared from synthetic and natural polymers. Beads containing ion-exchange and chelate-forming groups, or hydrophobic and affinity ligands have been

© Springer Nature Singapore Pte Ltd. 2018
K. Saito et al., *Innovative Polymeric Adsorbents*,
https://doi.org/10.1007/978-981-10-8563-5_3

developed depending on the type and concentration of target ions and foreign ions. For example, a target protein may be captured because of an electrostatic interaction: By properly adjusting the pH of the solution, a protein with a positive or negative charge in biological fluids may be collected using cation- or anion-exchange resin, respectively.

We refer to a bead with functionality as a functional bead. Components dissolved in a liquid, e.g., ions, are separated by flowing the liquid through a functional bead-packed bed; an appropriate pressure applied to the bed with a pump is required to cause the liquid to flow through the bed in an upward or downward direction. Target ions are transported by inducing them to flow through the interstices between beads and diffuse from the outer surface of a bead to the interior to reach the functional groups or ligands.

Smaller functional beads packed into a bed provide a higher overall adsorption rate of target ions and a higher operational pressure of the pump. In other words, conventional bead-packed beds give rise to a trade-off in that decreasing the diffusional mass-transfer resistance of ions increases the flow-through resistance of liquids. We must optimize materials and operations for separation to improve this trade-off.

3.1.2 Improving the Trade-off in Separations

Thus far, scientists and engineers have prepared materials used for separation and reaction involving ions and molecules, i.e., adsorbents and catalysts, respectively. Most conventional adsorbents and catalysts are shaped like beads and granules. For example, most commercially available organic ion exchangers are prepared from styrene-divinylbenzene copolymer beads. Granular-activated carbons are produced by activating pulverized natural products such as nut shells with chemicals or steam.

These beads and granules are packed into a bed for industrial use. Beads with smaller diameters enhance the mass transfer of target ions and molecules into their interior; however, a bed packed with smaller beads requires a higher pressure for fluids to flow through the bed at a practical flow rate. Similarly, some catalyst-packed beds have the problem that overall catalytic reaction rates are diffusion-limited in the case of a fast intrinsic catalytic reaction; therefore, smaller catalysts are required to raise the total activity of such beds.

To overcome this issue, chemical engineers have improved the structure of adsorbents and optimized the operating conditions of adsorptive beds. Instead of diffusion-limited mass transfer, convection-aided mass transfer is favorable for separation using the adsorbents. Afeyan et al. [1] in 1990 proposed "perfusion chromatography" of proteins using beads with a bimodal pore structure. Perfusion beads consist of throughpores and diffusive pores; a protein solution partially penetrates through the throughpores. On the other hand, Brandt et al. [2] in 1988 proposed "membrane chromatography" of proteins using modified porous hollow-fiber membranes. Protein solutions were forced to permeate through pores rimmed by affinity ligands.

These two breakthrough methods for purifying proteins are ascribed to sophisticated porous materials used, the pores of which enable convective flow of protein solutions. We learned that there was a need to design porous polymeric materials functioning in the convective mode. From this viewpoint, radiation-induced graft polymerization is promising in that it is readily applicable to the modification of existing porous polymeric materials.

3.1.3   Strategy for Revolution in the Form of Polymeric Adsorbents

Schemes for the preparation of polymeric adsorbents are shown in Fig. 3.1. First, a porous polyethylene hollow-fiber membrane used industrially as a microfiltration membrane with pore diameters of 0.1–0.5 μm was modified into various adsorbents for metals and proteins and used in the convective or permeation mode. Second, a porous polyethylene sheet with a pore diameter of 0.5 μm was adopted as a trunk polymer and modified into different adsorbents for analytical use by radiation-induced graft polymerization.

Third, commercially available nylon-6 fiber with a pore diameter of 35 μm was used as a starting polymer to obtain fibrous adsorbents. A fiber may be thought of as a strand of small beads with adjustable packing density, which enable the selection

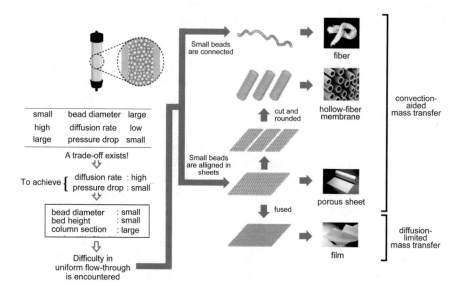

Fig. 3.1 Strategy for innovative polymeric adsorbents

of the flow-through resistance of a fiber-packed bed. Fourth, rolling up a modified nonwoven fabric with a spacer is an alternative for packing fibers into a bed. A commercially available nonwoven fabric made of polyethylene/polypropylene with spaces between fibers of approximately 100 μm was used as a trunk polymer.

In contrast, for diffusion dialysis and electrodialysis, convective flow is not permitted. Diffusion flux driven by the concentration gradient across the thickness of the membrane is essential for the improvement of dialyzer performance. Conventional methods of preparing ion-exchange membranes are tedious and time-consuming. We succeeded in simplifying the preparation of these membranes with their conventional performance retained. Commercially available nonporous films with thicknesses of 30–50 μm were adopted as the trunk polymer for radiation-induced graft polymerization.

3.2 Metal Ion Collection with Porous Hollow-Fiber Membranes

3.2.1 Selection of MF Membrane as a Starting Material

Various nonporous hollow-fiber membranes have been developed for manufacturing high-performance artificial kidneys. Hollow fibers are advantageous in that their outer surface area per volume is much larger than that of flat sheets. Therefore, a module consisting of a bundle of nonporous hollow-fiber membranes is being used in dialysis for the therapy of kidney disease. Undesirable solutes in blood are transferred from the blood side to the dialysate side through a nonporous hollow-fiber membrane because of transmembrane concentration gradients of the solutes.

Porous hollow-fiber membranes with pore sizes less than 1 μm have been developed for MF. MF membranes can screen out microorganisms and colloidal particles suspended in water while allowing the permeation of minerals dissolved in water, for example magnesium and calcium ions. Such MF membranes have been incorporated into devices for purifying drinking water and into equipment for producing large amounts of ultrapure water.

The types of polymeric MF membranes include polyethylene, polypropylene, and poly-N-vinylidene difluoride. For example, Asahi Kasei Chemicals Co. is manufacturing polyethylene MF membranes in the hollow-fiber form. Their pore sizes and thicknesses range from 0.1 to 0.5 μm and from 1 to 2 mm, respectively. We have thus far selected a polyethylene porous hollow-fiber membrane as a trunk polymer for grafting because the ratio of thickness to pore diameter of a modified porous hollow-fiber membrane is sufficiently large to enable overall adsorption with negligible mass-transfer resistance. The pores are rimmed by a graft chain containing various moieties capable of selectively collecting ions and molecules as a solution permeates through the pores.

An epoxy-group-containing vinyl monomer, glycidyl methacrylate (GMA), in methanol as a solvent was readily graft-polymerized onto an electron-beam-irradiated polyethylene porous hollow-fiber membrane. The degree of GMA grafting is defined as the weight increases percentage. At 100% GMA grafting, the poly-GMA graft chain was appended uniformly across the membrane thickness. A degree of GMA grafting of 142% obtainable at a reaction time of 20 min is equivalent to an epoxy-group density of 10 mol per kg of the trunk polymer. Subsequently, 50% molar conversion of the epoxy group into an iminodiacetate group [$-N(CH_3COOH)_2$] provides 5 mol/kg the trunk polymer or 2 mol/kg product.

3.2.2 Chelating Porous Hollow-Fiber Membranes

Membrane chromatography was proposed by Brandt et al. [2] for high-speed purification of proteins based on affinity interactions. The reason for high speed is that the time required for a protein to diffuse to the ligand immobilized on the surface of a pore in a porous hollow-fiber membrane is much shorter than the residence time of a protein solution as it passes through a pore across the membrane. That is, the diffusive time of a protein in the radial direction of a pore is negligible compared with the residence time of the protein solution during convective flow in the axial direction. This favorable mass-transfer mode also applies to the selective or specific collection of metal ions. We prepared chelating porous hollow-fiber membranes capable of selectively collecting metal ions at high speed by radiation-induced graft polymerization of GMA as a precursor monomer followed by introduction of chelating or chelate-forming groups via epoxy-ring opening.

The chelating groups introduced into the graft chain include iminodiacetate groups [3–6] for heavy metal ions such as copper, cobalt, and nickel ions, ethylenediamine [7] or guanine [8] moieties for palladium ions, the N-methylglucamine moiety [9, 10] for antimony, and triethanolamine units [11–13] for germanium compounds. These chelating groups were readily introduced into the poly-GMA graft chain to establish chelating group densities higher than those of commercially available chelating resin beads. For example, an iminodiacetate-type chelating porous hollow-fiber membrane had a chelating group density of 3 mol/kg H-form product, which was twofold that of a commercially available chelating resin, DIAION CR-11 (Mitsubishi Chemicals Co.).

Almost all chelating porous hollow-fiber membranes had three important features: high rates of adsorption and elution, high adsorption capacity at equilibrium, and high chemical durability under severe elution conditions. First, high adsorption rates are ascribed to convective transport of metal ions through the pores. Second, high chelating group density contributes to high adsorption capacity. The chelating group density is adjustable by determining the degree of GMA grafting and the molar conversion of epoxy groups into chelating groups. Third, the ester groups on the poly-GMA graft chain remaining after the introduction of the chelating groups are stable. These features enable their functionalization with various reagents under severe reaction conditions and the repeated elution of adsorbed metal ions with mineral acids such as 1 M hydrochloric acid.

3.2.3   Ideal Breakthrough Curve of Modified Porous Hollow-Fiber Membranes

A metal-ion-containing solution was fed to the inner surface of a chelating porous hollow-fiber membrane and forced to permeate radially outward through the pores at a constant flow rate using a syringe pump (Fig. 3.2). The effluent penetrating the outer surface of a hollow fiber was continuously collected in fraction vials. The metal ion concentration in each fraction was measured to generate a breakthrough curve (BTC), i.e., metal ion concentration in the effluent versus effluent volume. Here, effluent volume means not the volume of each fraction but the total volume of accumulated effluent. The flow rate of the solution varied in the range of one order of magnitude within the allowable mechanical strength of the hollow fiber.

The ordinate and abscissa of BTCs are the dimensionless metal ion concentration and effluent volume, respectively. The dimensionless metal ion concentration is the ratio of metal ion concentration in the effluent to that in the feed, whereas the dimensionless effluent volume (DEV) is obtained by dividing the effluent volume by the membrane volume. Here, the membrane volume is calculated as $\pi(d_o^2-d_i^2)L/4$, where d_o and d_i are the outer and inner diameters, respectively, and L is the length of the hollow fiber. This dimensionless expression is convenient for the comparison of the performance of hollow fibers with different thicknesses and lengths.

BTCs of an iminodiacetate-type chelating porous hollow-fiber membrane (IDA fiber) for cobalt ions are shown in Fig. 3.3 [5]. The flow rate of cobalt chloride ($CoCl_2$) solution was varied by changing permeation pressure over a range from 0.025 to 0.1 MPa. That is, the residence time of the solution moving through a pore across the IDA fiber was in the range from 3.45 to 0.86 s. The BTCs overlapped irrespective of the flow rate or residence time of the solution: A higher flow rate of $CoCl_2$ solution led to a higher overall adsorption rate of cobalt ions. This can be

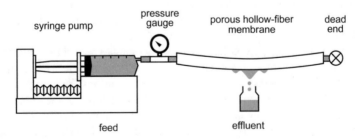

Fig. 3.2 Experimental apparatus for determination of flux and breakthrough curves during liquid permeation through modified porous hollow-fiber membranes

explained by considering that the diffusional mass-transfer resistance of cobalt ions to IDA groups is negligible and that the intrinsic chelate formation reaction is instantaneous. Because the overall adsorption rate of metal ions is governed by the flow rate of the solution as the feed, an ideal operation is achievable using chelating porous hollow-fiber membranes. In the case of using conventional adsorbents, e.g., chelating and ion-exchange beads, an increasing flow rate of the solution through the bead-packed bed results in an unfavorable BTC shape.

To quantify the merits of chelating porous hollow-fiber membranes, dynamic binding capacity (DBC) and equilibrium binding capacity (EBC) are defined from BTCs as follows:

$$DBC = \int_{0}^{V_B} (C_0 - C)dV/W \tag{3.1}$$

$$EBC = \int_{0}^{V_E} C_0 - CdV/W, \tag{3.2}$$

where C_0 and C are the metal ion concentrations in the feed and effluent, respectively; V and W are the effluent volume and mass of the IDA fiber, respectively; and V_B and V_E are the effluent volumes at which C reaches 0.1 C_0 and C_0, respectively. The fact that DBC remains constant at various flow rates of the solution reflects the

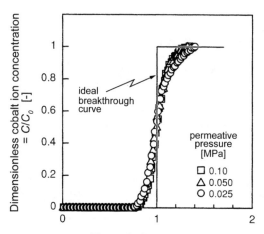

Fig. 3.3 Dimensionless breakthrough curves of chelating porous hollow-fiber membranes for cobalt ions at various permeative pressures of cobalt chloride solution. Reprinted with permission from Ref. [5]. Copyright 1992 American Chemical Society

rapid uptake of metal ions during permeation. Reasonably, EBC remains constant, irrespective of the flow rate.

3.2.4   Requirements for Rapid Uptake of Metal Ions

Metal ion collection during the permeation of a target solution through a chelating porous hollow-fiber membrane proceeds via the following four steps: (1) liquid-film diffusion of metal ions from the bulk of the solution to the inner surface of the hollow fiber aided by convective flow, (2) radial diffusion of metal ions through pores driven by the concentration gradient in the solution, (3) chelate formation by metal ions with chelating groups on the graft chain, and (4) graft-phase diffusion of metal ions along the chain driven by the gradient in the amount of metal ions adsorbed. Of the four steps, when the two diffusional steps (2) and (4) and the intrinsic reaction step (3) are fast, the overall collection rate of metal ions by the hollow fiber is governed by the first step: The flow rate of the feed solution determines the overall adsorption rate; that is, the residence time of the solution across the membrane determines the overall adsorption rate. This process is favorable for adsorption, which is referred to as a convection-limited or convection–aided process. This advantage is ascribed to the ratio of membrane thickness to pore diameter of approximately 10,000: The shortest pore length is 1 mm, and the largest pore diameter is 0.1 μm, as illustrated in Fig. 3.4.

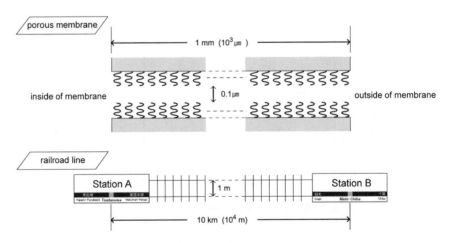

Fig. 3.4 Illustration of dimensions of modified porous hollow-fiber membrane

3.2.5   Direct Measurement of Amount of Metal Ions Adsorbed onto Hollow Fibers

Along with the determination of BTCs, the distribution of cobalt ions adsorbed across the IDA fiber was measured by X-ray microanalysis (XMA). Figure 3.5 shows that the amount of cobalt ions adsorbed is proportional to the area calculated from the line profile of XMA. The area as a function of effluent volume agreed well with the amount of adsorbed cobalt ions calculated from the BTCs. As the effluent volume increased, the line profile of the IDA fiber was almost rectangular with respect to the baseline. The sides of the rectangle shifted from the inner surface of the hollow fiber at x = 0 to the outer surface at x = 1 while retaining the same height.

A cobalt chloride ($CoCl_2$) solution was fed to the inner surface of the IDA fiber. Cobalt ions were instantaneously captured by the IDA groups on the graft chain until the amount of cobalt ions adsorbed equilibrated with the feed concentration. Therefore, as the effluent volume increased, maximum value of the profile shifted from the inner surface to the outer surface.

3.2.6   Rapid Uptake of Noble Metal Ions

A chelating porous hollow-fiber membrane capable of removing trace amounts of metal ions dissolved in ultrapure water may be applied to the recovery of noble

Fig. 3.5 Distribution of amount of cobalt bound across thickness of iminodiacetate-type porous hollow-fiber membrane

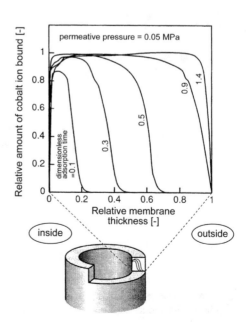

metal ions such as platinum (Pt), palladium (Pd), and rhodium (Rh) ions. The platinum compound $PtCl_2(NH_3)_2$, cisplatin, is an antitumor drug: Cisplatin cross-links nucleic acid bases of DNA to suppress cell proliferation.

We expected that nucleic acid bases would capture platinum group metals (PGMs). Of the four nucleic acid bases, adenine (A), guanine (G), and cytosine (C) can be readily added to the epoxy groups of the polymer chain grafted onto a porous hollow-fiber membrane, as shown in Fig. 3.6. The molar conversion of the epoxy groups into a nucleic acid base moiety (adenine) was in the range from 11 to 27%. The density of an immobilized nucleic acid base was in the range from 0.46 to 1.1 mol/kg of the hollow fiber. These densities were relatively higher than those of chelating beads.

A palladium chloride solution, the concentration of which was 100 mg-Pd/L of 1 mol/L hydrochloric acid, was forced to permeate through an adenine-immobilized porous hollow-fiber membrane (ADE fiber). The BTCs overlapped irrespective of the residence time of the palladium solution, which was in the range from 3 to 18 s (Fig. 3.7) [8]. The principle of negligible diffusional mass-transfer resistance of target ions to the chelating groups on the graft chain was also observed in this combination. As long as the intrinsic chelate formation is rapid, this principle is effective.

At pH 0, the ADE fiber exhibited higher binding capacity for palladium than other adsorbents. Palladium ions adsorbed onto this hollow fiber were quantitatively

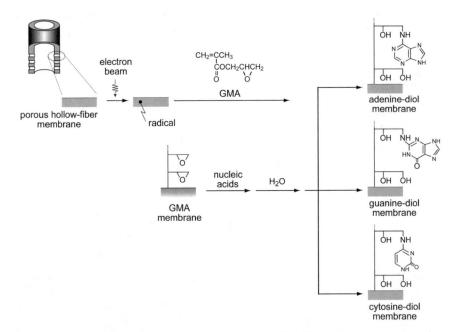

Fig. 3.6 Preparation scheme for porous hollow-fiber membrane with immobilized nucleic acid. Reprinted with permission from Ref. [8]. Copyright 2008 Elsevier

Fig. 3.7 Breakthrough curves of adenine-immobilized porous hollow-fiber membranes for palladium ions at various residence times. Reprinted with permission from Ref. [8]. Copyright 2008 Elsevier

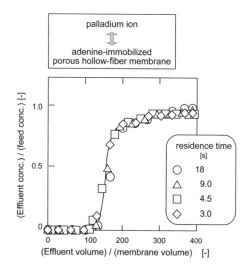

eluted with 1 M hydrochloric acid. In addition, cycles of adsorption and elution were repeated with no degradation of performance.

3.2.7   Recovery of Germanium Oxide [11]

Polyethylene terephthalate (PET) is a product of the condensation of ethylene glycol and terephthalic acid with water as a by-product. Germanium oxide (GeO_2) and antimony oxide (Sb_2O_3) have been used as homogeneous catalysts for this polymerization; therefore, a trace amount of the catalyst is contained in PET products such as transparent bottles and film.

Japan does not produce germanium. All germanium consumed in Japan is imported. To recover germanium oxide from various fluids, we immobilized tri-ethanolamine moieties on polymer chains grafted onto porous hollow-fiber membranes. Triethanolamine captures germanium oxide via covalent binding with the elimination of water, as shown in Fig. 3.8a. This structure is referred to as an atrane (germatrane for the combination of three hydroxyl groups with germanium). Upon treatment with an acid, germatrane hydrolyzes to release germanium oxide and water. These captures and releases may be regarded as adsorption and elution, respectively.

Iminodiethanol [$NH(CH_2CH_2OH)_2$] was added to the epoxy groups of graft chains of the porous hollow-fiber membrane (Fig. 3.8b). The resultant hollow fiber is referred to as an IDE fiber. When an aqueous solution of germanium oxide (GeO_2) permeated through the pores of the IDE fiber at various flow rates, the BTCs overlapped irrespective of the flow rate or residence time of the solution (Fig. 3.9). In this case, the higher the flow rate of the GeO_2 solution, the higher the

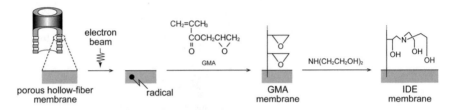

(a) Complex formation between triethanolamine and GeO₂

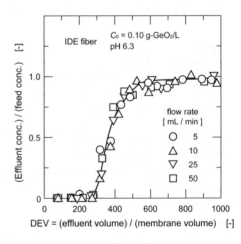

(b) Immobilization of iminodiethanol to graft chain

Fig. 3.8 Atrane structure and immobilization of iminodiethanol on graft chain of porous hollow-fiber membrane. Reprinted with permission from Ref. [11]. Copyright 2000 Elsevier

Fig. 3.9 Breakthrough curves of iminodiethanol-immobilized porous hollow-fiber membranes for germanium oxide at various flow rates. Reprinted with permission from Ref. [11]. Copyright 2000 Elsevier

binding rate of GeO₂. Furthermore, GeO₂ bound to the hollow fiber was quantitatively eluted by permeation with 1 M hydrochloric acid. A stable repeated use of capture and release was demonstrated.

3.2.8 Removal of Boron

Boron is an essential element for plants; however, it is harmful to animals. Boron has been used to disinfect water and to control cockroaches. N-Methylglucamine, which contains four hydroxyl groups, was added to a GMA-grafted porous hollow-fiber membrane, resulting in the formation of five hydroxyl groups on the membrane (Fig. 3.10). The resultant polyol-group-containing porous hollow-fiber membrane is referred to as an NMG fiber. Three hydroxyl groups of the five react with boron species in aqueous media. The BTCs overlapped irrespective of the flow rate of the $B(OH)_3$ solution. The absence of dependence of BTCs on flow rate in the permeation mode was favorable for the removal of hazardous ions. However, the porous hollow-fiber membrane as a starting material is costly, so we use a commercially available nylon fiber as an alternative polymer [14].

3.2.9 Limit of Almost Zero of Diffusional Mass-Transfer Resistance

Ideal separation procedures were achieved using porous hollow-fiber membranes with immobilized graft chains containing iminodiacetate groups, adenine, iminodiethanol, and N-methylglucamine. For these membranes, BTCs overlapped irrespective of the flow rate of the target solution because of the negligible diffusional mass-transfer resistance of target ions to functional moieties. Radiation-induced graft polymerization was effective in appending functional groups and ligands uniformly across porous hollow-fiber membranes, the thickness of which was approximately 1 mm.

Modified porous hollow-fiber membranes do not always exhibit ideal adsorption characteristics. When the following inequality is established, the lack of dependence of BTC on residence time of the solution no longer holds:

$$t_r < t_d + t_i + t_g, \tag{3.3}$$

where t_r is the residence time of the solution and t_d, t_i, and t_g are the radial diffusion time of metal ions in the pore, the time required for chelate formation of metal ions

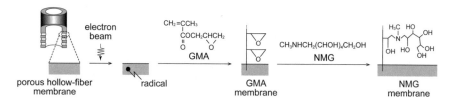

Fig. 3.10 Addition of N-methylglucamine to epoxy ring of graft chain

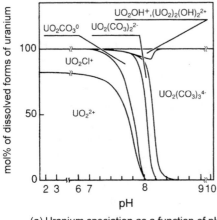

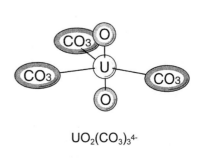

(a) Uranium speciation as a function of pH

(b) Predominant form of uranyl ionic species dissolved in seawater

Fig. 3.11 Dissolved forms of uranium

with the chelating group, and the diffusion time of metal ions in the graft phase, respectively.

Two cases satisfy this inequality (3.3). First, t_r on the left side of Eq. (3.3) decreases as the flow rate of the solution increases; however, a higher transmembrane pressure is required for a higher flow rate. The shortest t_r is determined by the upper limit of the allowable mechanical strength of the hollow fiber. Second, slow chelate formation causes t_i to increase. This second case was observed in the combination of amidoxime groups on the graft chain with uranyl species dissolved in seawater, i.e., uranyl tricarbonate ions $[UO_2(CO_3)_3^{4-}]$, as shown in Fig. 3.11.

Amidoxime groups capture uranyl tricarbonate ions to form new complexes while eliminating carbonate ions from the uranyl complex ions. Uranium was collected from seawater, allowing prefiltered seawater to permeate through the pores of amidoxime-group-containing porous hollow-fiber membranes. On the

Fig. 3.12 Overall adsorption rate of uranium during permeation of seawater through amidoxime-type porous hollow-fiber membranes

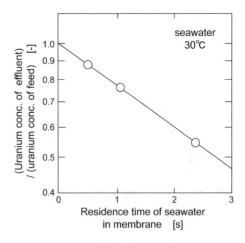

assumption that the overall adsorption rate of uranium is proportional to the uranium concentration in the liquid, the uranium concentration of the effluent decreased exponentially with residence time (Fig. 3.12) [15]. Relatively low concentrations of uranyl tricarbonate ions (3 μg-U/L or 1.3×10^{-8} M) and the relatively complicated exchange of ligands surrounding the uranyl ions result in intrinsic chelate-formation-limited adsorption.

3.3   Metal Ion Collection with Extractant-Impregnated Porous Hollow-Fiber Membranes

3.3.1   Separation of Metal Ions with Extractants

Extractants have been synthesized for the selective collection of target metal ions such as rare-earth metal ions, noble metal ions, and radioactive nuclides. They can be categorized into acidic, neutral, or basic extractants (Fig. 3.13). Representative extractants contain several long alkyl chains; for example, HDEHP and Aliquat 336 possess two branched octyl and three straight octyl chains, respectively. The long alkyl chains of extractants ensure their high solubility in organic solvents.

Solvent extraction methods have been utilized for the recovery of rare-earth and noble metal ions on both analytical and industrial scales. However, the use of solvents as an organic phase and the loss of extractants into an aqueous phase are the drawbacks of these methods. To overcome these drawbacks, solid-phase extraction (SPE) was proposed for analytical use [16]. Also, acidic extractant-impregnated beads, LEWATIT^{TM}, manufactured by Lanxess, are commercially available for large-scale or industrial solid-phase extraction.

Organic solvents are replaced with long alkyl moieties introduced into the graft chain, such as octyl and dodecyl groups, that interact with the hydrophobic parts of the extractants. That is, solvent-free extraction is achieved. The amount of extractants leaked is reduced by adjusting the strength of the hydrophobic interaction between the extractant and the hydrophobic ligands on the graft chains. Thus, we have prepared extractant-impregnated porous hollow-fiber membranes.

3.3.2   Impregnation of Acidic Extractants into Graft Chains [17–20]

HDEHP [bis(2-ethylhexyl)phosphate], an acidic extractant, was impregnated into polymer chains grafted onto porous hollow-fiber membranes. Alkyl amine $C_nH_{2n+1}NH_2$ with n in the range from 2 to 18 was added to the epoxy groups of a GMA graft chain to attract the two branched hydrophobic parts (2-ethylhexyl groups), as shown in Fig. 3.14. The simultaneous introduction of a secondary amino group (–NH–) enhances the impregnation of HDEHP owing to electrostatic

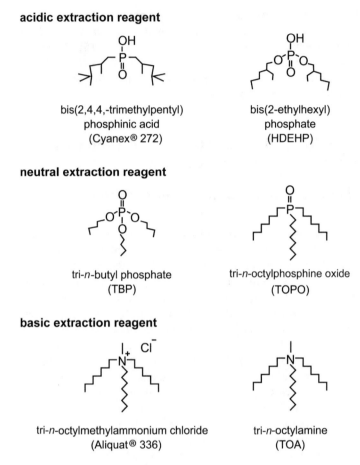

acidic extraction reagent

bis(2,4,4,-trimethylpentyl)
phosphinic acid
(Cyanex® 272)

bis(2-ethylhexyl)
phosphate
(HDEHP)

neutral extraction reagent

tri-*n*-butyl phosphate
(TBP)

tri-*n*-octylphosphine oxide
(TOPO)

basic extraction reagent

tri-*n*-octylmethylammonium chloride
(Aliquat® 336)

tri-*n*-octylamine
(TOA)

Fig. 3.13 Molecular structures of representative extractants

interactions between the positive charge of –NH- and the negative charge of the HDEHP anion.

An aqueous solution of yttrium oxide (Y_2O_3) dissolved in 0.01 mol/L nitric acid was forced to permeate through an HDEHP-impregnated porous hollow-fiber membrane. Dodecylamino groups ($C_{12}H_{25}NH–$) provided the highest adsorption capacity for yttrium ions. In addition, the breakthrough curves of the HDEHP-impregnated porous hollow-fiber membranes for yttrium ions overlapped irrespective of the flow rate of yttrium solution (Fig. 3.15); an ideal separation using extractant-impregnated porous hollow-fiber membranes was demonstrated in solid-phase extraction.

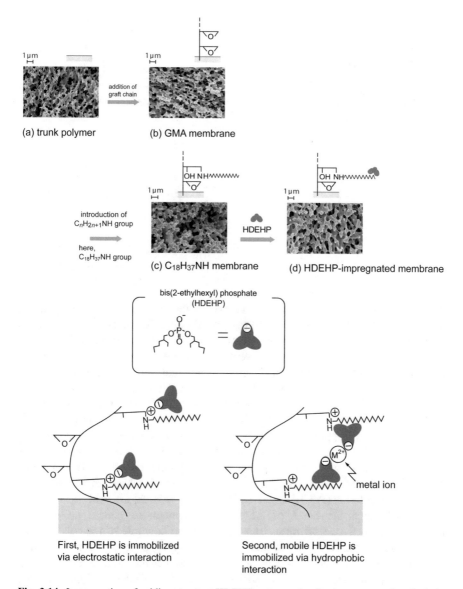

Fig. 3.14 Impregnation of acidic extractant HDEHP onto octadecylamino groups of graft chain

3.3.3 Impregnation of Basic Extractants into Graft Chains [21, 22]

The representative basic extractant Aliquat 336 [tri-*n*-octylmethylammonium chloride] was impregnated into polymer chains grafted onto porous hollow-fiber membranes. An alkylamino-group-containing graft chain does not serve as an

Fig. 3.15 Breakthrough
curves of
HDEHP-impregnated porous
hollow-fiber membrane for
yttrium ions at various flow
rates. Reprinted with
permission from Ref. [19].
Copyright 2005 Elsevier

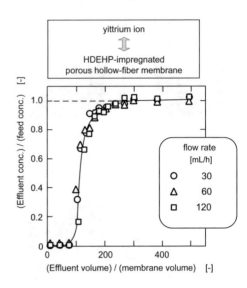

interaction pair because the positive charge of the protonated secondary amino
group (–NH-) repels the positive charge of Aliquat 336; consequently, a novel
scheme for the impregnation of basic extractants was devised (Fig. 3.16).

Two kinds of functional groups were successively introduced into the epoxy
group of the GMA polymer chains grafted onto the porous hollow-fiber membranes: the carboxyl group (–COOH) and the octadecyl group (–C$_{18}$H$_{37}$). The
former group allows the graft chain to swell, attracting Aliquat 336 on the basis of
electrostatic interactions. The latter group enhances the impregnation of Aliquat 336
on the basis of hydrophobic interactions. As a result, Aliquat 336 was incorporated
at a density of 1.2 mmol per g of GMA-grafted fiber. Here, Aliquat 336 molecules
are not linked to the graft chain via covalent binding but are mobile and do not leak
out of the graft chain into the aqueous phase during solid-phase extraction.

Palladium chloride (PdCl$_2$) dissolved in 1 mol/L hydrochloric acid was forced to
permeate through the pores of Aliquat 336-impregnated porous hollow-fiber
membranes at various flow rates. Irrespective of the flow rate of the solution, the
breakthrough curves in the permeation mode overlapped, because the time required
for ions to diffuse into the extractant-impregnated graft chain and to form a complex
with the extractant is much shorter than the residence time of the
target-ion-containing solution as it moves through the pores of the hollow fiber.

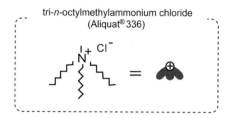

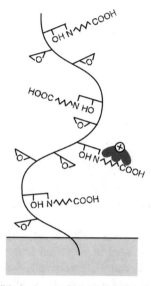

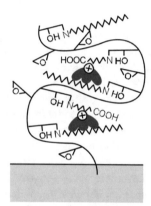

Possible leakage of Aliquat® 336 induced by electrostatic repulsion between

ammonium group of Aliquat® 336 and 6-aminohexacarboxyl group of graft chain

Enhanced impregnation of Aliquat® 336 by hydrophobic interaction between

hydrophobic group of Aliquat® 336 and octadecyl group of the graft chains

Fig. 3.16 Impregnation of basic extractant Aliquat® 336 into hydrophobic graft chains. Reprinted with permission from Ref. [21]. Copyright 2006 Elsevier

3.3.4 Impregnation of Neutral Extractants into Graft Chains [23]

Impregnation of neutral extractants as well as acidic and basic extractants is required to extend the application of porous hollow-fiber membranes to the collection of valuable metal species. A two-layer structure of functional groups along the length of graft chains was utilized to impregnate neutral extractants into the graft chains. Diol and octadecyl groups were introduced into the upper and lower parts of the graft chains, respectively. The epoxy group was reacted with water in 0.5 M H_2SO_4 before it was reacted with octadecane thiol ($C_{18}H_{37}SH$) (Fig. 3.17).

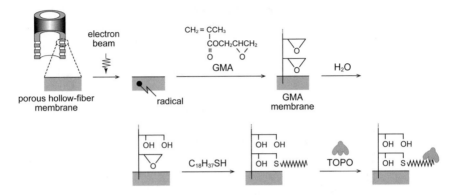

Fig. 3.17 Preparation scheme for TOPO-impregnated porous hollow-fiber membranes

The neutral extractant, tri-*n*-octylphosphine oxide (TOPO), was impregnated into the octadecyl group in the lower part in methanol. Subsequently, the hollow fiber was thoroughly washed with water. Thus, TOPO was capped by the diol group in the upper part of the graft chain. This two-layer structure prevents the leakage of TOPO from the graft chain to the external solution (Fig. 3.18). When $Bi(NO_3)_3$ dissolved in 1 mol/L hydrochloric acid permeated through the pores of a TOPO-impregnated porous hollow-fiber membrane, Bi^{3+} was captured by the graft chain and TOPO was not detected in the effluent.

Four representative extractants, i.e., HDEHP, Cyanex 272 [bis(2,4,4-trimethyl pentyl)phosphinic acid], Aliquat 336, and TOPO, were impregnated onto porous hollow-fiber membranes to capture ions of Y, Zn, Pd, and Bi, respectively, during the permeation of their solutions through the pores of the hollow fibers. In all combinations, a rapid uptake of target ions was demonstrated because of negligible mass-transfer resistances and instantaneous complex formation. Furthermore, quantitative elution of the target ions with appropriate eluents was also demonstrated. The successful repeated use of the extractant-impregnated porous hollow-fiber membranes was achieved.

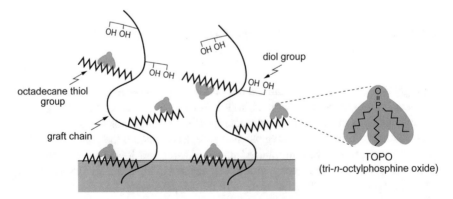

Fig. 3.18 Possible structure of TOPO impregnated into diol- and octadecane thiol-containing graft chains

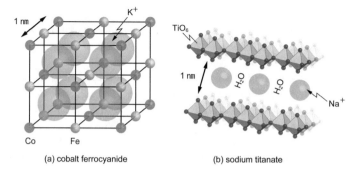

(a) cobalt ferrocyanide   (b) sodium titanate

Fig. 3.19 Crystal structures of cobalt ferrocyanide and sodium titanate

   Immiscible water and oil phases were vigorously mixed to ensure a large inter-facial area and to enhance mass transfer, which may be replaced by the convective flow of aqueous solutions through pores rimmed by extractant- impregnated hydrophobic graft chains at a high density. For example, the acid extractant HDEHP was impregnated at a density as high as 2.0 mol/kg. Loss of extractants caused by the formation of microemulsions from the oil phase was prevented by the interaction of the hydrophobic moieties of the extractant with the hydrophobic ligands on the graft chain. For example, the phosphor was not detected in the effluent that passed through the HDEHP-impregnated porous hollow-fiber membrane. Rapid uptake and elution were demonstrated in the permeation mode using an extractant-impregnated porous hollow-fiber membrane as a support in solid-phase extraction, which corresponds to a rapid forward and backward extraction carried out in a liquid–liquid extraction.

   We have introduced ion-exchange and chelate-forming groups into the graft chains for selective adsorption of metal ions. Furthermore, we have immobilized extractants onto hydrophobic graft chains for highly selective adsorption of rare-earth metal ions. Since the Fukushima Daiichi nuclear disaster on March 11, 2011, adsorbents for the removal of radionuclides such as cesium-137 and strontium-90 from contaminated water have been developed by a number of universities, research institutes, and private companies. Our group focused on insoluble cobalt ferrocyanide and sodium titanate, which specifically incorporate cesium and strontium ions, respectively, as shown in Fig. 3.19. From the viewpoint of direct contact of adsorbent with contaminated water and subsequent easy recovery of the adsorbents, fibers were adopted as trunk polymers for grafting instead of porous hollow-fiber membranes. Immobilization procedures for inorganic compounds and their removal performance of the modified fibers are detailed in Chap. 4.

3.4   Protein Capture with Porous Hollow-Fiber Membranes

Proteins with higher molecular masses than heavy metal ions have lower diffusivity, resulting in lower diffusive flux. Therefore, rapid uptake of proteins in the perme-ation mode across a modified porous hollow-fiber membrane is achievable if the diffusional mass-transfer resistance is negligible owing to the convective flux of the protein solution through the pores. In addition to the rapid uptake of proteins, multilayer binding of proteins via multiple points along charged graft chain is observable, as described in the following.

Target proteins may be captured on the basis of electrostatic, hydrophobic, and affinity interactions. Electrostatic interactions are also referred to as anion- or cation-exchange interactions. Affinity interactions include pseudoaffinity interactions. Functional groups and ligands that exhibit interactions are attached to the polymer chains grafted onto porous hollow-fiber membranes. The resultant porous hollow-fiber membranes are applicable to the recovery of useful proteins and the removal of undesirable proteins. Although ion-exchange groups have been introduced into the polymer brushes and roots of graft chains, proteins are exclusively accessible to the ion-exchange groups on the polymer brushes because of their exclusion based on their size from the polyethylene matrix incorporating the polymer roots.

3.4.1   Concerns About Polyethylene-Based Adsorbents for Protein Purification

To selectively capture a target protein, the following interactions between proteins and adsorbents have been adopted: ion-exchange or electrostatic, hydrophobic, and affinity interactions. Here, affinity ligands are categorized into biospecific ligands such as Protein A and pseudobiospecific ligands such as immobilized nickel. When the ligands corresponding to these interactions are immobilized on the polymer chains grafted onto porous hollow-fiber membranes by radiation-induced graft polymerization, high-rate or high-capacity capturing of proteins is achievable.

In the beginning of our development of polyethylene-based adsorbents, experts on protein purification criticized our strategy to introduce ligands into a polymer chain grafted onto a polyethylene (PE) porous hollow-fiber membrane because proteins would nonselectively adsorb onto the PE interface. They assumed that the PE-based adsorbents would not be suitable for protein purification. Nonselective adsorption of proteins onto an adsorbent surface induces their deformation, resulting in incomplete elution of proteins from the surface; thus far, agarose-based adsorbents such as Sepharose beads supplied by GE Healthcare Co. have been used to purify proteins. Agarose, a natural polysaccharide, is relatively hydrophilic, whereas PE is relatively hydrophobic.

3.4.2 Hydrophilization of Polyethylene Surfaces with Graft Chains [24–26]

Grafting of an epoxy-group-containing vinyl monomer, glycidyl methacrylate (GMA), onto a PE substrate is convenient for attaching various ligands and hydrophilic moieties. We assessed ways to minimize nonselective adsorption of proteins by hydrophilization of grafted poly-GMA chains. A GMA-grafted porous hollow-fiber membrane was reacted with 0.5 M sulfuric acid to quantitatively convert epoxy groups into diols, i.e., two adjacent alcoholic hydroxyl groups (Fig. 3.20). The amounts of bovine gamma globulin (BGG) bound to the diol-containing porous hollow-fiber membranes derived from GMA-grafted porous hollow-fiber membranes with the degree of GMA grafting reaching up to 200% are shown in Fig. 3.21 as a function of diol group density. At a diol group density higher than 7 mmol/g, the amount of BGG bound was reduced to approximately one-sixth of that of the starting PE porous hollow-fiber membrane.

The degree of hydrophilization of the PE surface was also evaluated from the pure water flux (PWF) of a porous hollow-fiber membrane that was previously dried under reduced pressure. PWF was measured by feeding pure water to the inner surface of a hollow fiber at a prescribed pressure, e.g., 0.05 MPa. The flow rate of pure water penetrating the outer surface of the hollow fiber was measured, as described in Fig. 1.13 of Chap. 1. The PWF determined by dividing the measured flow rate by the inner surface area of the dried hollow fiber was compared with that of the hollow fiber that was previously immersed in methanol for a sufficient time to fill the pores with methanol and that was then immersed in water.

As shown in Fig. 3.22, the PWFs of the dried hollow fiber and the pre-methanol-treated hollow fiber overlapped at diol group densities higher than

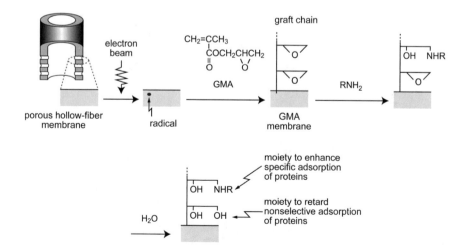

Fig. 3.20 Successive introduction of two kinds of functional groups into graft chains

Fig. 3.21 Binding of globulin onto hydrophilic-group-containing graft chains

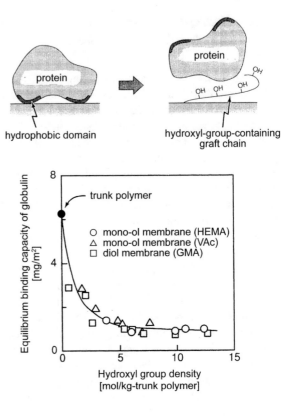

Fig. 3.22 Pure water flux of diol fiber versus its hydroxyl group density. Reprinted with permission from Ref. [26]. Copyright 2008 John Wiley & Sons Ltd.

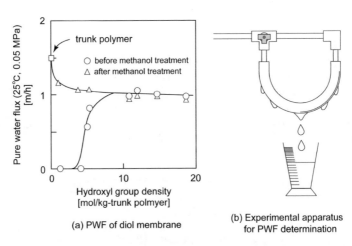

6-7 mmol/g. This indicates that the PE surface is covered with hydrophilic graft chains forming a self-wettable surface. Complete hydrophilization of the diol-group-containing polymer chain is achieved at this density from the viewpoint of surface wettability.

3.4.3 Immobilization of Betaine [27]

Serum is a biological fluid that contains various proteins, and it flows in blood vessels, where nonselective adsorption of proteins to vessel surfaces does not occur. On the basis of the understanding of the structure of cell membranes, Professor K. Ishihara of the University of Tokyo developed a novel vinyl mono-mer,2-methacryloyloxyethyl phosphorylcholine(MPC), which mimics the phospholipid bilayer of cells [28]. MPC is an amphoteric vinyl monomer containing both a phosphoric acid group and a quaternary ammonium chloride group. Poly-MPC is reported to form a surface inert to the nonselective binding of proteins.

Presently, MPC as a phosphobetaine is an expensive monomer, whereas sulfobetaine and carboxybetaine are relatively economical. We immobilized N,N-dimethyl-γ-aminobutyric acid as a carboxybetaine onto a GMA-grafted porous hollow-fiber membrane in an effort to decrease the amount of proteins adsorbed. The amount of lysozyme adsorbed onto the fiber was reduced to approximately one-tenth of that adsorbed onto the diol fiber. A complete minimization of nonselective binding of proteins onto the polymeric surface is not required for polymeric adsorbents designed for protein purification. We selected not carboxybetaine but the diol or a 2-hydroxyethylamino group as the group to coexist with ion-exchange groups and biospecific ligands.

3.4.4 Anion-Exchange Porous Hollow-Fiber Membranes [29–43]

Ion-exchange reactions occur on the basis of electrostatic interactions between two different ions. Therefore, ion exchangers containing positively charged groups can exchange anions, whereas those containing negatively charged groups can exchange cations: The exchangers are referred to as anion and cation exchangers, respectively. In this book, we describe organic ion exchangers derived from existing polymers and inorganic ion exchangers immobilized onto organic polymers.

The principle of ion exchange is applicable to protein purification. Of various ion-exchange groups, diethylamino and sulfonic acid groups are commonly used anion- and cation-exchange groups, respectively (Fig. 3.23). Proteins possess an isoelectric point (pI). For example, Tris-HCl, phosphate, and acetate buffers, the pHs of which reach up to 8, render albumins (pI 5) anionic macromolecular species in the buffer solutions. On the other hand, lysozymes (pIs 10–11) dissolve as cationic macromolecular species in buffer solutions. In fundamental studies, bovine serum albumin (BSA) and hen egg lysozyme (HEL) have been used to evaluate the protein binding performance of anion and cation exchangers, respectively.

Diethylamine [DEA, $NH(C_2H_5)_2$] and 2-hydroxyethylamine (HEA, [$NH_2C_2H_4OH$]) were successively reacted with a GMA-grafted porous hollow-fiber membrane. The resultant weakly basic anion-exchange porous hollow-fiber membrane is referred to as a DEA-HEA fiber. When the flow rate of a BSA solution (0.02 mol/L Tris-HCl buffer, pH 8.0) through the pores of the DEA-HEA fiber ranged from 20 to 100 mL/h, i.e., an average residence time from 165 to 33 s, the obtained dimensionless breakthrough curves overlapped irrespective of flow rate, as shown in Fig. 3.24. A higher flow rate of a protein solution through the pores of an ion-exchange porous hollow-fiber membrane led to a higher overall adsorption rate of the protein to the polymer brushes expanding from the pore surface. This favorable result occurs because the diffusional mass-transfer resistance of BSA to the DEA group [$-N(C_2H_5)_2$] is negligible owing to the convective flow of the BSA solution.

Additionally, BSA adsorbed onto the DEA-HEA fiber was quantitatively eluted by permeating 0.5 M sodium chloride across the hollow fiber. After conditioning the hollow fiber with the buffer, the breakthrough curve was satisfactorily reproduced; the HEA group [$-NHC_2H_4OH$] coexisting with the DEA group was found to be effective for reducing nonselective adsorption of the protein onto the hollow fiber.

Ion-exchange porous hollow-fiber membranes are advantageous over conventional ion-exchange beads in that a linear scale-up is possible: The size of the separation space capable of recovering proteins is constant irrespective of the number of hollow fibers used. A linear scale-up using a module consisting of eight anion-exchange porous hollow-fiber membranes was experimentally demonstrated. Generally, the height increase in a bead-packed bed does not permit a linear scale-up to predict the performance of protein adsorption.

Fig. 3.23 Two ion-exchange groups capable of capturing proteins

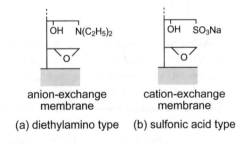

anion-exchange   cation-exchange
membrane   membrane

(a) diethylamino type   (b) sulfonic acid type

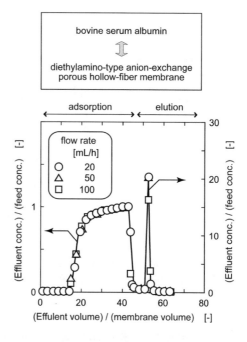

Fig. 3.24 Breakthrough and elution curves of diethylamino-type anion-exchange porous hollow-fiber membrane for bovine serum albumin. Reprinted with permission from Ref. [29]. Copyright 1995 Elsevier

3.4.5   Cation-Exchange Porous Hollow-Fiber Membranes [44–51]

The sulfonic acid group ($-SO_3H$) as a cation-exchange group was introduced into a GMA-grafted porous hollow-fiber membrane using sodium sulfite (SS, $NaHSO_3$) as a sulfonation reagent. The remaining epoxy group was converted into a diol group. Here, a diol group coexisting with a sulfonic acid group behaves as a hydrophilic group to reduce the nonselective binding of proteins. The resultant strongly acidic cation-exchange porous hollow-fiber membrane is referred to as an SS-diol fiber. As the molar conversion of the epoxy group into the SS groups increased, the pure water flux drastically decreased. The degree of extension of the strongly acidic cation-exchange graft chain of the SS-diol fiber is much higher than that of the weakly basic anion-exchange graft chain of the DEA-HEA fiber; at an identical molar conversion of epoxy groups into ion-exchange groups, the SS-diol fiber exhibited a much lower flux than the DEA-HEA fiber.

The flux and protein binding capacity of the SS-diol fiber at a molar conversion of 10% were evaluated for the HEL solution (0.02 mol/L phosphate buffer, pH 6). Irrespective of the flow rate of the HEL solution, which ranged from 10 to 100 mL/h, i.e., an average residence time in the range from 100 to 10 s, the dimensionless breakthrough curves overlapped (Fig. 3.25). The favorable characteristic of negligible diffusional mass-transfer resistance was observed in the SS-diol fiber as well as in the DEA-HEA fiber. In addition, HEL adsorbed onto the hollow fiber was eluted

Fig. 3.25 Breakthrough
curves of sulfonic acid-type
cation-exchange porous
hollow-fiber membrane for
lysozyme at various flow
rates. Reprinted with
permission from Ref. [45].
Copyright 2008 John Wiley &
Sons Ltd.

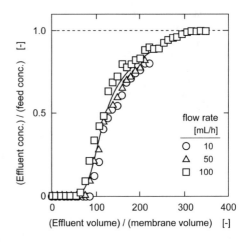

with 0.5 M NaCl and achieved an elution percentage of 100%. After conditioning the
hollow fiber with buffer, a reproducible breakthrough curve was observed.

3.4.6 Degree of Multilayer Binding [52]

Protein binding in multilayers to polymer brushes provides an adsorbent with a high
capacity. This phenomenon is favorable for protein purification and enzyme
immobilization. For example, the sulfonic acid-type cation-exchange porous
hollow-fiber membrane exhibited an equilibrium binding capacity for lysozyme of
0.42 g per g of hollow fiber [46]. This capacity is approximately tenfold higher than
those of conventional adsorbents in bead form.

The porous hollow-fiber membrane has a spongelike pore network with an
average pore size of 0.4 μm and a porosity of approximately 70%. The specific
surface area of the hollow fiber was approximately 10 m^2/g. The radiation-induced
graft polymerization of an epoxy-group-containing vinyl monomer and subsequent
introduction of ion-exchange groups produced a charged polymer brush extending
from the pore surface of the hollow fiber.

A more precise description of the charged polymer brush is as follows: First,
irradiation of electron beams or gamma rays onto a PE porous hollow-fiber
membrane generates radicals over the entire volume of the hollow fiber consisting
of the PE matrix and pore. Second, the polymer chain grows from the radicals in
both the crystalline and amorphous domains to form graft chains. For a convenient
simple understanding of the graft chain, a graft chain is classified into a polymer
root or brush according to the formation site. The polymer root is the part of the
graft chain embedded in the matrix, and the polymer brush is the part of the graft
chain extending from the pore surface toward the pore interior. Third, the
ion-exchange groups are introduced into the epoxy groups of the graft chain,

Fig. 3.26 Two extreme adsorption orientations of albumin onto a solid surface

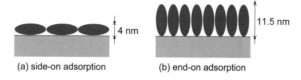

(a) side-on adsorption   (b) end-on adsorption

i.e., the polymer root and brush. In this reaction, a partial introduction of ion-exchange groups along a graft chain is possible by appropriate selection of solvents for the reaction. A brush with an ion-exchange group or any ionizable group is referred to as a charged polymer brush.

Proteins that are products of the condensation of amphoteric amino acids possess charges depending on the pH of their solutions. A protein with a negative or positive charge is accessible to the ends of a polymer brush with the opposite charge and may bind to the polymer brush at multiple points through electrostatic interactions. A bound protein can diffuse along a polymer brush driven by the gradient of the amount of protein bound. This diffusion phenomenon is referred to as a graft-chain phase diffusion. Graft-chain phase diffusion through a graft chain occurs in the direction of membrane thickness.

When the gradient of the amount of protein bound reaches zero, a binding equilibrium is attained. The equilibrium binding capacity of a protein to the polymer brush is convertible into the degree of multilayer binding defined as the ratio of equilibrium binding capacity to that of theoretical monolayer adsorption capacity. Here, the theoretical monolayer adsorption capacity was calculated by assuming that protein molecules cover the pore surface in a monolayer with two extreme orientations as follows (Fig. 3.26):

$$q_t = (\mathrm{Mr}\ a_v)/(a\ N_A), \qquad (3.4)$$

where Mr and N_A are the molecular mass of a protein and Avogadro's number, respectively, a_v is the specific surface area of the hollow fiber, and a is the apparent area occupied by a protein molecule.

Proteins are entangled by the polymer brush via multiple points through electrostatic interactions. In addition, when different proteins are used as probes for the polymer brush, different degrees of multilayer binding are obtained. Therefore, the degree of multilayer binding is a rough estimate of the length of the polymer brush. For example, an equilibrium binding capacity for lysozyme of 0.42 g per g of sulfonic acid cation-exchange porous hollow-fiber membrane corresponded to a degree of multilayer binding of 38 [42].

3.4.7 Immobilization of Taurine [53]

The sulfonic acid group (–SO$_3$H) is a representative of strongly acidic cation-exchange groups; therefore, a SO$_3$H-group-containing graft chain extends

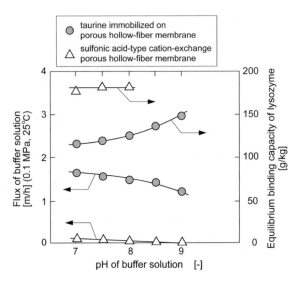

Fig. 3.27 Flux of buffer solution and equilibrium binding capacity of lysozyme of taurine immobilized on porous hollow-fiber membrane as a function of pH of buffer solution. Reprinted with permission from Ref. [53]. Copyright 2005 Elsevier

into the bulk because of the mutual electrostatic repulsions of the SO_3H groups. To mitigate the degree of extension of the graft chain, we introduced taurine ($NH_2CH_2CH_2SO_3H$) into the GMA graft chain by partial neutralization of SO_3H groups with imino groups ($-NH-$). As shown in Fig. 3.27, the pure water flux of a taurine-immobilized porous hollow-fiber membrane with a degree of GMA grafting of 150% and a molar conversion of the epoxy group into the taurine moiety of 25% was 30-fold higher than that of a SO_3H-group-containing porous hollow-fiber membrane with an identical degree of GMA grafting and a molar conversion of the epoxy group into the SO_3H group of only 6%. On the other hand, the equilibrium binding capacity of the former hollow fiber for lysozyme decreased by 30% compared with that of the latter hollow fiber. The immobilization of taurine moieties on the polymer chains grafted onto the porous hollow-fiber membranes prevented a drastic reduction in the water permeability while retaining the protein adsorptivity of the membranes.

3.4.8 Immobilization of Hydrophobic Ligands [54–57]

Phenyl and long-chain alkyl groups as hydrophobic ligands were introduced into GMA-grafted porous hollow-fiber membranes. For example, phenol and butyl thiol or octyl thiol were used to append hydrophobic ligands to the graft chain. The resultant hydrophobic porous hollow-fiber membranes are applicable to protein purification. Proteins dissolved in a high-ionic-strength solution, such as 4 M ammonium sulfate [$(NH_4)_2SO_4$], bind to the hydrophobic ligands of polymer brushes. Subsequently, bound proteins may be eluted with an $(NH_4)_2SO_4$–free solution.

Fig. 3.28 Breakthrough curves of phenyl-type porous hollow-fiber membrane for albumin at various permeation pressures. Reprinted with permission from Ref. [56]. Copyright 1997 Elsevier

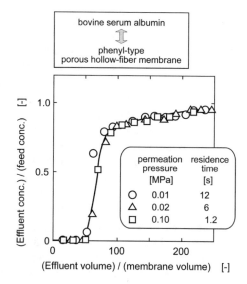

BSA dissolved in 4 M $(NH_4)_2SO_4$ was fed to a phenyl-group-containing porous hollow-fiber membrane, Ph fiber, in the permeation mode at various flow rates to generate breakthrough curves. The obtained breakthrough curves overlapped irrespective of the residence time of the BSA solution across the Ph fiber (Fig. 3.28). BSA was captured by polymer brushes with a negligible diffusional mass-transfer resistance, as observed in chelating and ion-exchange porous hollow-fiber membranes.

The ion-exchange polymer brushes function through mutual electrostatic repulsion of the ion-exchange groups expanded to hold BSA in multilayers. On the other hand, the hydrophobic ligands with almost no charge do not allow the graft chains to swell, resulting in the monolayer binding of BSA onto the pore surface of the Ph fiber. When a cycle of adsorption and elution using the Ph fiber was repeated, no quantitative elution was achieved; as the number of repeated uses increased, the binding capacity of the Ph fiber decreased. Prior to adsorption, washing with 1 M sodium hydroxide was effective in recovering the initial binding capacity of the Ph fiber for BSA.

3.4.9   Immobilization of Affinity Ligands [58–60]

The affinity interactive force is defined as a comprehensive force between a ligand and a ligate, where "comprehensive" means the sum of electrostatic and hydrophobic interactions, metal chelation, and hydrogen bonding. An antigen/antibody or antibody/antigen pair is representative of affinity ligand/ligate pairs. Affinity ligands are categorized into two groups: biospecific and pseudobiospecific ligands. Biospecific ligands such as monoclonal and polyclonal antibodies capture their corresponding proteins specifically or highly selectively, and then the bound proteins are eluted with strong eluents that may denature the proteins. On the other hand, the pseudobiospecific affinity interactions are useful because elution is carried out under conditions milder than those for biospecific affinity interactions.

To preconcentrate 17β-estradiol dissolved in surface water, such as river water, 1000-fold, its polyclonal antibody was immobilized on the GMA-grafted porous hollow-fiber membrane via an epoxy-ring opening reaction. 17β-Estradiol solution (1 ng/L) as a model feed was forced to permeate radially outward through the pores rimmed by the polymer brushes with the immobilized polyclonal anti-17β-estradiol antibody. The obtained breakthrough curves overlapped irrespective of flow rate (Fig. 3.29). This characteristic is favorable for the preconcentration of analytes because analysts are unaware of the flow rate of a sample solution through a hollow fiber used as an adsorbent. The "$t_D \ll t_R$ principle" is effective for this antigen/ antibody combination because the antigen, 17β-estradiol, with its small molecular mass (Mr 272), ensures an instantaneous interaction with its polyclonal antibody.

17β-Estradiol that bound to polymer brushes was quantitatively eluted with methanol. However, when the hollow-fiber membrane was washed with water for reuse, the binding capacity of the hollow fiber for 17β-estradiol in the second cycle decreased to 44% of the initial value (Fig. 3.30). Methanol significantly damages the conformation of the polyclonal antibody immobilized to the polymer brushes.

Fig. 3.29 Breakthrough curves of antibody to 17β-estradiol immobilized on porous hollow-fiber membrane at various flow rates. Reprinted with permission from Ref. [58]. Copyright 2002 American Chemical Society

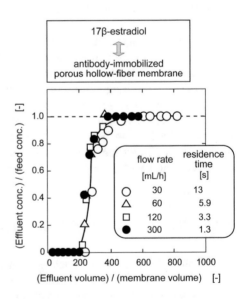

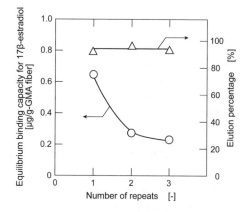

Fig. 3.30 Equilibrium binding capacity of antibody to 17β-estradiol immobilized on porous hollow-fiber membrane and elution percentage versus number of repeated adsorptions and elutions. Reprinted with permission from Ref. [58]. Copyright 2002 American Chemical Society

3.4.10   Immobilization of Pseudoaffinity Ligands [61–68]

The pseudobiospecific ligands include immobilized metal affinity (IMA) ligands. For example, nickel ions immobilized by an iminodiacetate group, [-N(CH$_2$COOH)$_2$], selectively capture histidine (His)-tagged proteins. Silver ions immobilized by a sulfonic acid group selectively bind double-bond-containing molecules, e.g., polyunsaturated fatty acids (PUFAs). Affinity chromatography using immobilized metal ions is referred to as immobilized metal affinity chromatography (IMAC). Silver chromatography is used when silver ions are the metal ions in IMAC.

We prepared a porous hollow-fiber membrane containing a copper-ion-IDA ligand (Fig. 3.31) and, using BSA as a model protein originally containing His residues, demonstrated its merit over conventional affinity beads because of the $t_D \ll t_R$ principle. However, copper ions were partially eluted by BSA. Instead of

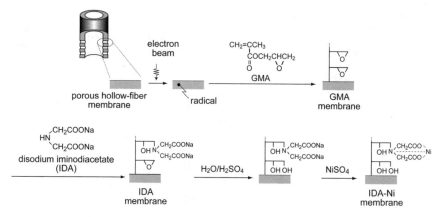

Fig. 3.31 Preparation scheme for immobilized metal affinity porous hollow-fiber membranes

copper ions, nickel ions were immobilized by the IDA group of the graft chain, and a solution of a His-tagged fusion protein as a real solution permeated through the hollow fiber. The protein bound to a monolayer was quantitatively eluted by permeation with a 50 mM imidazole solution. A comparison of the lanes obtained by SDS-PAGE between the feed during adsorption and the effluent during elution demonstrated that the His-tagged fusion protein was purified using the nickel-ion-immobilized porous hollow-fiber membrane in the permeation mode.

3.4.11 Purification of DHA [69, 70]

PUFAs, such as docosahexaenoic acid (DHA) and eicosapentaenoic acid (EPA) used in pharmaceuticals, are contained in the oil of fish such as bonito and tuna. The DHA ethyl ester from bonito oil ethyl esters is purified using a silver-ion-containing metal affinity porous hollow-fiber membrane. This feed is a multicomponent solution containing DHA and other ethyl esters (DHA-Et and Other-Ets, respectively). The breakthrough and subsequent elution curves are shown in Fig. 3.32, where bonito oil ethyl ester containing 66 wt% DHA-Et was dissolved in methanol to give a feed concentration of 5 g/L. In Fig. 3.32, the abscissa and ordinate are dimensionless effluent volume and concentration, respectively, defined as follows.

$$\text{Dimensionless effluent volume (DEV)} \\ = (\text{effluent volume})/(\text{membrane volume excluding the lumen}) \tag{3.5}$$

$$\text{Dimensionless concentration } (C/C_0) \\ = (\text{effluent concentration})/(\text{feed concentration}) \tag{3.6}$$

Fig. 3.32 Breakthrough curves of ethyl esters in tuna oil in porous hollow-fiber membranes with immobilized silver ions. Reprinted with permission from Ref. [70]. Copyright 2001 John Wiley & Sons Ltd.

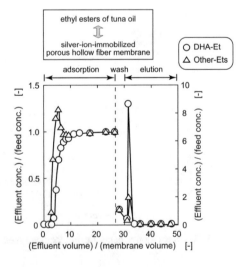

Neither DHA-Et nor Other-Ets appeared up to DEV = 2. At DEV = 3, Other-Ets concentration reached a breakthrough point and then sharply increased. DHA-Et was detected at DEV = 4 and Other-Ets exceeded the feed concentration at DEV = 5. Then, the Other-Ets concentration reached a peak of $C/C_0 = 1.24$ and decreased to the feed value. In contrast, the DHA-Et concentration increased until equilibrium was attained.

This "roll-up" phenomenon was observed similarly in the breakthrough curves of the chelating and ion-exchange porous hollow-fiber membranes for binary metal ions and proteins, respectively, caused by displacement adsorption based on the order of affinity: DHA ethyl ester, with a higher affinity for the immobilized silver ions, repelled other ethyl esters during the continuous permeation of the feed through the pores rimmed by silver-ion-immobilized graft chains.

Breakthrough curves of iminodiacetate porous hollow-fiber membranes for cobalt and copper ions (Co^{2+} and Cu^{2+}) at various permeation pressures are shown in Fig. 3.33 [71]. That there is no dependence of the shape of the breakthrough curve on the permeation pressure, i.e., residence time, indicates negligible diffusional mass-transfer resistance of the metal ions to the iminodiacetate groups immobilized on the graft chains. The amount profiles of Co^{2+} and Cu^{2+} sorbed in the hollow fiber as a function of effluent volume are shown in Fig. 3.34.

We now discuss the combined mass transfer of binary metal ions with chelate formation in the permeation mode. At an initial stage [(1) in Fig. 3.34], the effluent contained neither Co^{2+} nor Cu^{2+} ions; the Cu^{2+} ions, with a higher affinity for the IDA group, occupied the inner part (lumen side) of the hollow fiber, whereas the Co^{2+} ions preceded the Cu^{2+} ions and occupied more of the outer part (shell side). Here, the stability constants of Cu^{2+} and Co^{2+} ions as the iminodiacetate complexes are $10^{10.55}$ and $10^{6.95}$ [72], respectively, in 0.1 mol/L KCl at 303 K. At a second stage [(2) in Fig. 3.34], first, the Co^{2+} ions reached a breakthrough point; second, the Co^{2+} concentration in the effluent exceeded that of the feed; and third, the Cu^2 ions were

Fig. 3.33 Breakthrough curves of iminodiacetate-type chelating porous hollow-fiber membranes for Cu^{2+}-Co^{2+} binary system. Reprinted with permission from Ref. [71]. Copyright 1996 Elsevier

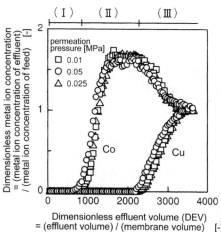

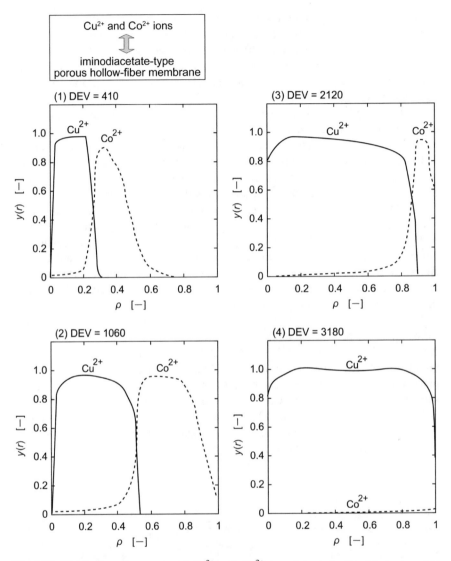

Fig. 3.34 Distribution of amounts of Cu^{2+} and Co^{2+} bound to iminodiacetate-type porous hollow-fiber membranes across membrane thickness. Reprinted with permission from Ref. [71]. Copyright 1996 Elsevier

detected in the effluent. The Cu^{2+} ions continued displacing the sorbed Co^{2+} ions toward the shell side. At a final stage [(3) in Fig. 3.34], almost all of the Co^{2+} ions were displaced by the Cu^{2+} ions; no additional amounts of the metals were sorbed during permeation. The displacement adsorption across the porous hollow-fiber membrane with immobilized silver ions is illustrated in Fig. 3.35: The high-affinity species (DHA-Et) displaces the low-affinity ones (Other-Ets).

As observed in the adsorption of binary metal ions onto the chelating porous hollow-fiber membrane, the displacement adsorption was observed during the continuous permeation of a binary protein solution (pH 8), which was a mixture of ovomucoid (OM: pI 4.1, Mr 28 kD) and ovotransferrin (OTf: pI 6.1, Mr 76.6 kD) through the pore of the anion-exchange porous hollow-fiber membrane (Fig. 3.36) [37]. Owing to the differences in pI and pH, OM with a higher affinity for the anion-exchange group repelled OTf with a lower affinity.

Displacement adsorption has also been examined in the ion-exchange bead bed or activated carbon-packed bed with a height of approximately 1 m. It is reasonable that the phenomenon is reproducible in the porous hollow-fiber membrane with a thickness of approximately 1 mm. Unlike the bead-packed bed, displacement adsorption is independent of the flow rate of a binary protein solution owing to negligible diffusional mass-transfer resistance of a protein to the ion-exchange group of polymer brushes and instantaneous intrinsic ion exchange.

The elution percentage of DHA-Et with acetonitrile was calculated to be 100%; the quantitative elution of DHA-Et was verified by the permeation of acetonitrile through the pores. In addition, repeated use of the porous hollow-fiber membrane

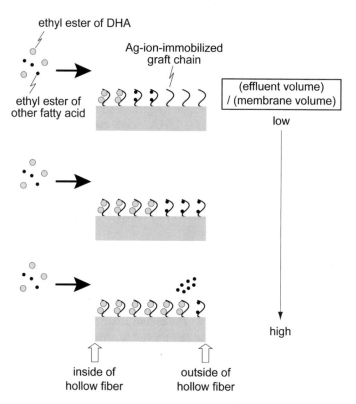

Fig. 3.35 Displacement of adsorbed DHA ethyl ester by other ethyl esters during permeation of bonito oil ethyl ester solution across porous hollow-fiber membranes with immobilized silver ions

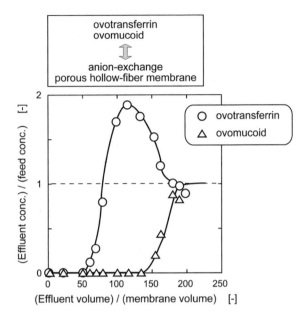

Fig. 3.36 Breakthrough curves of anion-exchange porous hollow-fiber membranes for ovotransferrin and ovomucoid

with immobilized silver ions for adsorption and elution revealed that the decrease in the amount of adsorbed DHA-Et was negligible after two cycles and no leakage of silver ions was detected. Currently, this porous membrane-based purification method for DHA ethyl ester from fish oil ethyl esters is inferior to the solvent extraction method using a silver nitrate ($AgNO_3$) solution on the basis of the criterion of the binding capacity for DHA ethyl ester. Solvent extraction is advantageous from the viewpoint of adsorption capacity because a saturated $AgNO_3$ solution contains silver ions at a concentration of 13 mol/L at 293 K.

3.5   Immobilized-Enzyme Reaction with Porous Hollow-Fiber Membranes

3.5.1   Multilayering Structure of Enzymes Bound to Graft Chains

Most enzymes are soluble in aqueous media and catalyze specific reactions under mild conditions, e.g., at ambient temperature and atmospheric pressure. Substrates are acted upon by enzymes in a homogeneous system and participate in reactions; however, from the viewpoint of the industrial use of enzymes, even after a complete enzymatic reaction, separation of products from enzymes is required to purify the products and reuse the enzymes. To promote the use of enzymatic reaction systems, enzymes have been immobilized by numerous researchers on various solid supports

Fig. 3.37 Multilayer binding of enzyme to polymer chains grafted onto pore surface. Reprinted with permission from Ref. [31]. Copyright 1995 Elsevier

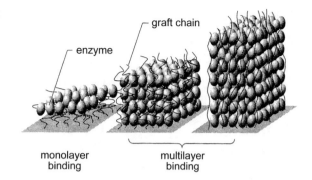

monolayer
binding

multilayer
binding

or matrices by various methods such as covalent binding, adsorption, and physical entrapment.

Ion-exchange graft chains hold enzymes as one class of proteins in multilayers, as illustrated in Fig. 3.37. An enzyme-multilayered structure can be achieved within the polymer brushes grafted onto porous hollow-fiber membranes. Furthermore, enzymes immobilized on hollow fibers at a high density have high activity because the substrates are fed with a negligible diffusional mass-transfer resistance to the immobilized enzymes. The total activity governed by an intrinsic enzymatic reaction or a substrate feed rate is superior to that determined by substrate diffusion. Enzymes immobilized on the porous hollow-fiber membranes are listed in Table 3.1.

3.5.2 Decomposition of Dextran into Cycloisomaltooligosaccharide [73–75]

Cycloisomaltooligosaccharide glucanotransferase (Mr 98 kDa) decomposes dextran, a straight-chain polysaccharide consisting of glucose monomers, into cycloisomaltooligosaccharide, a seven- to nine-membered cyclic glucose. Cycloisomaltooligo saccharide as a product of this enzymatic reaction functions as an inhibitor of glucosyltransferase contained in *Streptococcus mutans*. Glucosyltransferase GTF produces glucan with high adhesivity to *S. mutans* in our mouths, increasing the population density of *S. mutans*. Because S. mutans produces acid, this increase lowers the pH in the mouth causing cavities to form. Therefore, cycloisomaltooligosaccharide is a promising candidate for the prevention of formation of cavities. In addition, cycloisomaltooligosaccharide may be acceptable to humans because it is tasteless, odorless, and colorless.

Cycloisomaltooligosaccharide glucanotransferase was stacked in approximately five layers with the polymer brushes extending from the pore surface of an anion-exchange porous hollow-fiber membrane in the permeation mode. The binding of a negatively charged enzyme to an anion-exchange polymer brush is based on an

Table 3.1 Enzymes immobilized on graft chains

Enzyme	Substrate	Product
Cycloisomaltooligosaccharide synthase	Dextran	Cycloisomaltooligosaccharide
Collagenase	Gelatin	Tripeptide
α-amylase	Starch	Low molecular-weight starch
Urease	Urea	Ammonia and carbon dioxide
Ascorbic acid oxidase	Ascorbic acid and oxygen	Dehydroascorbic acid
Aminoacylase	DL-methionine	L-methionine
Dextran sucrase	Sucrose	Dextran

electrostatic interaction, because the pH of the enzyme solution is higher than the isoelectric point (pI) of the enzyme. In most cases, the pH suitable for enzyme binding does not coincide with the optimal pH for enzyme activity. To prevent enzyme leakage from the polymer brushes, cycloisomaltooligosaccharide glucan-otransferase molecules were enzymatically crosslinked with transglutaminase.

The conversion of dextran (average molecular mass: 43000) into cycloisomal-tooligosaccharide increased with an increasing degree of multilayering of cycloi-somaltooligosaccharide glucanotransferase. However, the conversion curve leveled off as the degree of enzyme multilayering increases, because a higher degree of enzyme multilayering into a polymer brush hinders the diffusion of dextran as a high-mass substrate deep into the enzyme-immobilized polymer brush.

3.5.3 Synthesis of Dextran from Sucrose [76]

Using a cycloisomaltooligosaccharide glucanotransferase-immobilized porous hollow-fiber membrane, the substrate dextran was decomposed into cycloisomal-tooligosaccharide. Furthermore, we attempted to produce dextran as a product from sucrose used as a substrate using a dextransucrase-immobilized porous hollow-fiber membrane.

Successive utilization of dextransucrase- and cycloisomaltooligosaccharide glucanotransferase-immobilized porous hollow-fiber membranes can produce cycloisomaltooligosaccharide from sucrose. First, an aqueous sucrose solution (0.1 mol/L acetate buffer, pH 5.5) was fed to the inner surface of dextransucrase-immobilized porous hollow-fiber membranes at a constant flow rate using a syringe pump. Initially, the effluent penetrated the outer surface of the hollow fiber without detection of dextran. Then, the pressure required to maintain a constant flow rate increased sharply (Fig. 3.38) and the pump stopped because it exceeded the allowable upper limit of applied pressure. This was a serious problem.

We observed a cross section of a hollow fiber by scanning electron microscopy. The inner surface of the hollow fiber was coated with a thin white layer (Fig. 3.39).

Fig. 3.38 Increase in permeation pressure with amount of dextran produced during the permeation of sucrose solution through the pores of DSase-immobilized porous hollow-fiber membrane. Reprinted with permission from Ref. [76]. Copyright 2004 John Wiley & Sons Ltd.

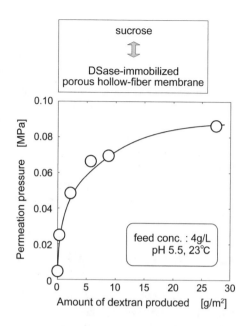

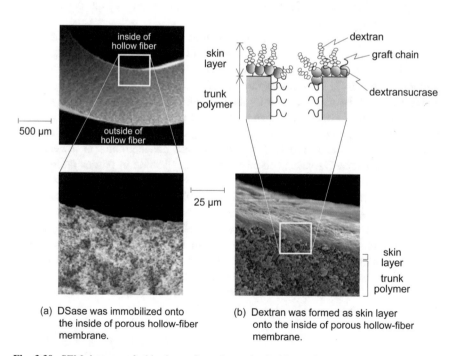

(a) DSase was immobilized onto the inside of porous hollow-fiber membrane.

(b) Dextran was formed as skin layer onto the inside of porous hollow-fiber membrane.

Fig. 3.39 SEM images of skin layer formed on the inside surface of porous hollow-fiber membrane with immobilized DSases. Reprinted with permission from Ref. [76]. Copyright 2004 John Wiley & Sons Ltd.

The dextran layer appeared to clog the pores, increasing the permeation pressure. The clogging of the pores of the microfiltration (MF) membrane caused by the enzymatic reaction is applicable to the pore-size tailoring of membranes: An MF membrane can be converted into ultrafiltration (UF) and reverse osmosis (RO) membranes under a controlled degree of enzymatic reaction capable of producing the thin layer of dextran.

Dextransucrase continues to capture the product at the active site to polymerize sucrose, as illustrated in Fig. 3.39. We observed that dextran produced by dextransucrase immobilized onto the polymer brushes clogs the pores of the microfiltration hollow-fiber membrane. This clogging has the potential to influence the pore size. Also, the surface coated with the thin layer of dextran may exhibit antifouling properties because of the hydrophilicity of dextran.

3.5.4 Decomposition of Urea [77]

Urease with a relatively high mass of 480 kDa decomposes urea to form ammonium ions and carbon dioxide. Urease was bound to a diethylamino-type anion-exchange porous hollow-fiber membrane at an amount of as much as 1.2 g per g of the hollow fiber. Immobilization of urease at a high density could have practical applications.

Urea solutions at high concentrations such as 4 M have been used to denature proteins. In contrast, a semipermeable dialysis film can be used to refold proteins. If a dialysis film bag containing proteins dissolved in a highly concentrated urea solution is immersed in a vessel that contains a large volume of water, urea is transported into the vessel driven by the concentration gradient of urea across the film, while water is transported from the vessel to the bag driven by the osmotic pressure difference across the film. This dialysis technique has disadvantages: Namely, a long time is required and the protein concentration is lowered. To overcome these disadvantages, we proposed the application of porous hollow-fiber membranes with immobilized urease to the removal of urea from protein solutions.

A urea solution at a concentration of 4 mol/L was forced to permeate through the pores of a porous hollow-fiber membrane with immobilized urease. The percentage decomposition of urea, defined as the ratio of effluent concentration to feed concentration expressed as a percent, is shown in Fig. 3.40 as a function of space velocity, defined as the ratio of flow rate to membrane volume. Up to a space velocity of 20 h^{-1}, a quantitative hydrolysis of urea was achieved. Urease immobilized on a graft chain and subsequently crosslinked works while being denatured by highly concentrated urea.

Fig. 3.40 Percentage hydrolysis of 4 M urea as a function of space velocity by porous hollow-fiber membrane with immobilized ureases. Reprinted with permission from Ref. [77]. Copyright 2003 John Wiley & Sons Ltd.

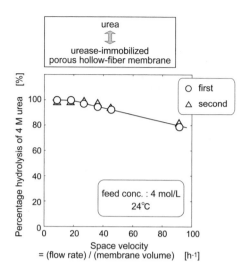

3.5.5 Kinetics of Enzyme-Immobilized Porous Hollow-Fiber Membranes [78–82]

Our next subject is the kinetics describing the permeation of a substrate solution through the pores of a porous hollow-fiber membrane with immobilized enzyme. First, the substrate solution is fed into the inner surface of the hollow fiber; second, during the permeation through the pores of the hollow fiber, the substrate reaches the active site of an enzyme immobilized on the polymer brush to carry out an enzymatic reaction.

The details of overall process are: (1) radial diffusion of the substrate toward the graft chain while flowing through the pores driven by the pressure difference, (2) the diffusion of the substrate into the graft chain, and (3) the intrinsic enzyme reaction at the active site of an immobilized enzyme. The observed or overall enzymatic reaction is the combined result of mass transfer and the intrinsic reaction.

Ascorbic acid (vitamin C) solutions at concentrations ranging from 0.025 to 0.10 mmol/L were permeated through the pores of a porous hollow-fiber membrane with immobilized ascorbic acid oxidase. The percentage decomposition of ascorbic acid is shown in Fig. 3.41 as a function of space velocity, defined as the ratio of the flow rate of the ascorbic acid solution to the volume of the membrane including the lumen. In this space velocity range, the percentage decomposition was constant at 100%: A quantitative decomposition was achieved. This favorable result can be explained by noting that the time required to pass through steps (1)–(3) is much shorter than the residence time of the substrate solution in the enzyme-containing porous hollow-fiber membrane. That is, the overall reaction is governed by the flow rate of the substrate solution and not by diffusion nor intrinsic reaction rate.

The flow rate of the substrate solution is limited by the pressure applied to the hollow fiber. Therefore, to improve the overall reaction efficiency, the amount of

Fig. 3.41 Percentage
decomposition of ascorbic
acid as a function of space
velocity in porous
hollow-fiber membrane with
immobilized ascorbic acid
oxidase

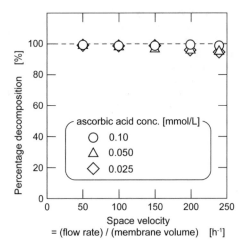

enzyme immobilized should be reduced. Enzymes immobilized on a polymer brush
may be resistant to the deterioration induced by conformational changes because the
enzymes bind to the brush at multiple points and crosslinked with transglutaminase.

3.6   Chiral Separations with Porous Hollow-Fiber Membranes

3.6.1   Immobilization of Albumin on Polymer Brushes [83–87]

One function of albumin dissolved in the blood is the transport of L-amino acids
throughout the body; therefore, albumin is a chiral selector capable of recognizing
L-amino acids. We have thus far evaluated the binding performance of proteins in
terms of rate and capacity using BSA, and we have demonstrated that BSA is
entangled in multilayers by the anion-exchange polymer brush of the anion-exchange
porous hollow-fiber membrane. The resultant BSA-multilayered structure provides a
novel domain favorable for chiral separation.

BSA was bound to a diethylamino-type anion-exchange porous hollow-fiber
membrane at pH 8.0 (20 mM Tris-HCl buffer). Subsequently, the bound BSA was
crosslinked with transglutaminase to prevent the leakage of BSA from the polymer
brush at a pH suitable for chiral separation and different from the pH of BSA binding.
Transglutaminase, which has been commercialized by Ajinomoto Co. for adjusting
the viscosity of fish meal via the crosslinking of fish proteins, is a practical reaction
and convenient enzyme to use because of the mild conditions required for its
crosslinking reaction. As a result, 80% of bound BSA was immobilized on the
polymer brush.

Fig. 3.42 Breakthrough curves for DL-Trp, D-Trp, and L-Trp in porous hollow-fiber membrane with immobilized BSA Reprinted with permission from Ref. [85]. Copyright 1999 Elsevier

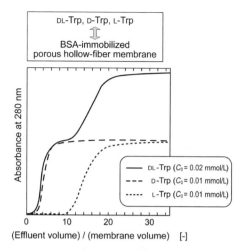

The chiral separation of DL-tryptophan was performed in both permeation and injection modes. The former and latter modes are described by breakthrough curves and chromatograms, respectively. In the permeation mode, the breakthrough curves for DL-tryptophan are shown in Fig. 3.42 along with those for D- and L-tryptophans, where the ordinate is the absorbance at 280 nm, which is linearly proportional to the tryptophan concentration. The breakthrough curve for DL-tryptophan exhibits a steplike shape and is equivalent to the sum of the breakthrough curves of D- and L-tryptophan. This result indicates that L-tryptophan exclusively interacts with the BSA molecules bound to the polymer brush.

An experimental apparatus for the injection mode of a chiral separation of DL-tryptophan with the porous hollow-fiber membrane with immobilized BSA as a stationary phase is illustrated in Fig. 3.43. A chromatogram in the injection mode is shown in Fig. 3.44. The retention time of the peak corresponding to D-tryptophan was equivalent to the residence time of the buffer as the mobile phase moving through the pores of the hollow fiber. The four-layered porous hollow-fiber

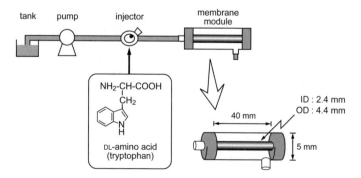

Fig. 3.43 Experimental apparatus for injection of DL-tryptophan

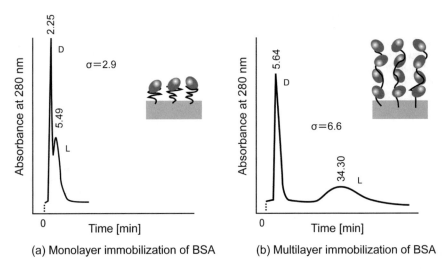

(a) Monolayer immobilization of BSA   (b) Multilayer immobilization of BSA

Fig. 3.44 Chromatogram of DL-tryptophan for injection mode in porous hollow-fiber membrane with immobilized BSA

membrane with immobilized BSA had a longer retention time than the monolayered porous hollow-fiber membrane with immobilized BSA. Here, the $t_D \ll t_R$ principle in both modes held true. Because the potential of albumin as a chiral selector is limited by its naturally occurring conformation, its applications are also limited. Chiral separation using immobilized proteins will be promising when proteins as the chiral selector for separating the targeted chiral compounds can be simply elaborated.

3.7 Porous Sheet

3.7.1 Solid-Phase Extraction

Solid-phase extraction (SPE) is a convenient method of preconcentrating target ions and molecules. The solid packed into an SPE cartridge works as an adsorbent on the basis of various interactions, e.g., electrostatic, hydrophobic, and affinity interactions. The adsorption of SPE cartridges must be evaluated in terms of adsorption rate and capacity. In addition, quantitative elution of the adsorbed target ions and molecules is required for SPE analysis.

Modes of packing the adsorbents into cartridges can be categorized into three groups: (1) bead-packed beds, (2) bead-/fiber-mixed beds, and (3) porous sheet-packed beds. The first mode is a conventional one, which is characterized by a trade-off between a higher capturing rate with smaller beads and a higher operating pressure with a higher flow rate. To improve this trade-off, the interstices among the smaller beads are interspaced by fibers in the second mode. The interstices generated by the fibers alleviate the operating pressure required for a high flow rate. The third mode employs a novel porous sheet as a packing material, whose pores are rimmed by modified graft chains.

We have adopted a commercially available porous sheet called "MAPS®" that is produced by INOAC Corporation. This polyethylene porous sheet has an average pore diameter of 1 μm and a porosity of 75% with a thickness of 2 mm. MAPS has been used industrially as a filter to remove suspensions from liquids and as an extruder pad to feed liquids at a desired rate.

The ratio of the pore length to the pore radius is about 10^3. This ratio is favorable for the SPE of target ions and molecules because negligible diffusional mass-transfer resistance is possible when an intrinsic binding reaction is instantaneous. The time required for the target ions and molecules to diffuse in the radial direction of the pores is much shorter than the residence time of the target solution as it moves through the pores.

Radiation-induced graft polymerization was applied to the modification of "MAPS." The modifications discussed in this book include the immobilization of functional groups and the impregnation of extractants into the MAPS. Radiation-induced graft polymerization was carried out to append epoxy-group-containing polymer chains over the entire MAPS uniformly across MAPS thickness: A uniform distribution of Cu^{2+} ions adsorbed across the thickness of an iminodiacetate-group-containing MAPS was produced.

3.7.2 Immobilization of Chelate-Forming Groups on MAPS [88–92]

Immobilization schemes for chelate-forming groups on MAPS are shown in Fig. 3.45. The epoxy groups on the poly-GMA chains grafted onto MAPS readily react with a chemical reagent, thereby immobilizing the iminodiacetate groups. These schemes consist of three steps: (1) electron beam irradiation, (2) graft polymerization of GMA, and (3) introduction of iminodiacetate groups.

The breakthrough curves of the iminodiacetate-group-containing MAPS (IDA-MAPS) for Cu^{2+} ions overlapped irrespective of the flow rate of the Cu^{2+} solution, which was in the range from 150 to 1500 mL/h. This range of flow rates corresponds to the range of the residence times from 9 to 0.9 s. A lack of the breakthrough curve on flow rate means that a higher flow rate of the Cu^{2+} solution provided a higher overall adsorption rate of Cu^{2+} into the IDA-MAPS. The same adsorption characteristics as the IDA-MAPS, which are favorable for separation procedures, were observed for a diethylamino-group-containing MAPS and a sulfonic-acid-group-containing MAPS.

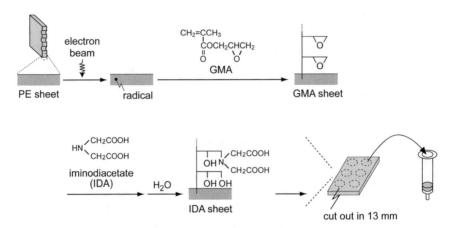

Fig. 3.45 Preparation scheme of iminodiacetate-type chelating porous sheet and its loading into commercially available SPE cartridges

3.7.3 Rapid Separation of Actinides in Spent Nuclear Fuels Using Anion-Exchange-Group-Containing MAPS [93–97]

The determination of the concentration and isotopic abundance of actinides in spent nuclear fuels is required for the cost-effective management of radioactive waste. The removal of interfering elements in spent nuclear fuel samples prior to measurements by inductively coupled plasma mass spectrometry (ICP-MS) enables an accurate determination of actinides. Thus far, numerous methods using the beds packed with anion-exchange and extractant-impregnated resins have been suggested. However, the resins in bead form have a disadvantage in that a longer time is required for the ionic species to diffuse into the interior of the resins with a larger diameter. That is, the adsorbents in bead form are characterized by having a trade-off between the diffusional mass-transfer rate of the analyte into a bead and the flow-through resistance of the sample solution across the bed. To overcome this drawback, a diethylamino-group-containing MAPS (DEA-MAPS) was employed to achieve a rapid separation of actinides.

U, Pu, and Am ions in the solutions of the spent nuclear samples were separated using a DEA-MAPS-packed bed in accordance with the procedure shown in Fig. 3.46. The leakage of U was less than 10^{-5} for a Pu fraction, which is favorable for the determination of trace amounts of ^{239}Pu by ICP-MS because the isobaric interference at a mass number of 239 caused by the production of ^{238}U^{1}H^{+} arising from the sample matrix is diminished. The results obtained by a novel method using the DEA-MAPS-packed bed agreed well with those obtained with a standard method using a conventional anion-exchange-bead (MCI-GEL, CA08Y)-packed bed. Trace amounts of Pu (0.34 pg) and U (38.16 pg) in the spent nuclear fuel

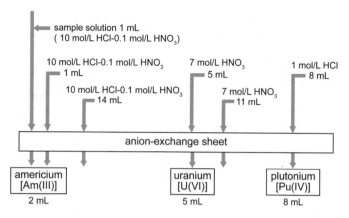

Fig. 3.46 Protocol for analysis of Am, U, and Pu

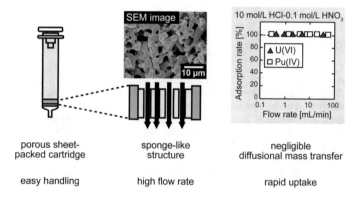

Fig. 3.47 Ion-exchange porous sheet capable of rapidly analyzing nuclides

sample were successfully determined using DEA-MAPS. The separation time, which was defined as the time required for a series of separation steps from conditioning to elution, was less than 1 h (Fig. 3.47).

3.7.4 Rapid Preconcentration of Rare-Earth Elements in Seawater Using Extractant-Impregnated MAPS [98, 99]

Rare-earth elements (REEs) have been used as a probe to elucidate geochemical processes in oceans. The concentrations of REEs in seawater have been determined by ICP-MS. Preconcentration is required for an accurate determination because their concentrations are equal to or below the detection limit (0.1 pg/g) of ICP-MS. Furthermore, matrix elements such as Na, K, Ca, and Mg should be removed to minimize damage to instruments.

Recently, extractant-impregnated SPE techniques based on the high selectivity of extractants for REEs have been applied to the preconcentration of REEs. For example, bis(2-ethylhexyl)phosphate (HDEHP) and 2-ethylhexyl dihydrogen phosphate (H_2MEHP) were impregnated into the hydrophobic substrate via hydrophobic interactions between the hydrophobic part of the extractants and the long alkyl chains of the substrate. However, a trade-off between mass-transfer rate and flow-through resistance exists in a resin-packed bed.

The HDEHP-impregnated MAPS (HDEHP-MAPS) offers a simpler procedure for the preconcentration of REEs than do the conventional adsorbents. The time required to preconcentrate REEs in a seawater sample using HDEHP-MAPS was threefold to 50-fold shorter than that using conventional adsorbents. The ratios of the values obtained using HDEHP-MAPS-packed beds to those reported by Zhang and Nozaki [100] ranged from 0.8 to 1.4.

References

1. N.B. Afeyan, N.F. Gordon, I. Mazsaroff, L. Varady, S.P. Fulton, Y.B. Yang, F.E. Regnier, Flow-through particles for the high-performance liquid chromatographic separation of biomolecules: perfusion chromatography. J. Chromatogr. **519**, 1–29 (1990)
2. S. Brandt, R.A. Goffe, S.B. Kessler, J.L. O'Connor, S.E. Zale, Membrane-based affinity technology for commercial scale purifications. Nat. Biotechnol. **6**, 779–782 (1988)
3. S. Tsuneda, K. Saito, S. Furusaki, T. Sugo, J. Okamoto, Metal collection using chelating hollow-fiber membrane. J. Membr. Sci. **58**, 221–234 (1991)
4. H. Yamagishi, K. Saito, S. Furusaki, T. Sugo, I. Ishigaki, Introduction of a high-density chelating group into a porous membrane without lowering the flux. Ind. Eng. Chem. Res. **30**, 2234–2237 (1991)
5. S. Konishi, K. Saito, S. Furusaki, T. Sugo, Sorption kinetics of cobalt in chelating porous membrane. Ind. Eng. Chem. Res. **31**, 2722–2727 (1992)
6. G. Li, S. Konishi, K. Saito, T. Sugo, High collection rate of Pd in hydrochloric acid medium using chelating microporous membrane. J. Membr. Sci. **95**, 63–69 (1994)
7. G. Li, S. Konishi, K. Saito, S. Furusaki, T. Sugo, K. Makuuchi, Collection of palladium using an ethylenediamine-immobilized chelating microporous membrane. Membrane (Maku) **20**, 224–228 (1995)
8. T. Yoshikawa, D. Umeno, K. Saito, T. Sugo, High-performance collection of palladium ions in acidic media using nucleic-acid-base-immobilized porous hollow-fiber membranes. J. Membr. Sci. **307**, 82–87 (2008)
9. S. Tsuneda, A. Hirata, M. Tamada, T. Sugo, High-speed recovery of antimony using chelating porous hollow-fiber membrane. J. Membr. Sci. **214**, 275–281 (2003)
10. T. Saito, H. Kawakita, K. Uezu, S. Tsuneda, A. Hirata, K. Saito, M. Tamada, T. Sugo, Structure of polyol-ligand-containing polymer brush on the porous membrane for antimony (III) binding. J. Membr. Sci. **236**, 65–71 (2004)
11. I. Ozawa, K. Saito, K. Sugita, K. Sato, M. Akiba, T. Sugo, High-speed recovery of germanium in a convection-aided mode using functional porous hollow-fiber membranes. J. Chromatogr. A **888**, 43–49 (2000)

12. H. Kim, M. Kim, I. Ozawa, K. Saito, K. Sugita, M. Tamada, T. Sugo, K. Sato, M. Akiba, K. Ichimura, Preparation of chelating porous membranes for the recovery of germanium and their adsorption characteristics. J. Ion Exchange **13**, 10–14 (2002)

13. T. Mochizuki, K. Saito, K. Sato, M. Akiba, T. Sugo, Recovery of p.t-CEtGeO using chelating porous membranes prepared with various compositions of dioxane/water solvent. J. Ion Exchange **18**, 68–74 (2007)

14. K. Ikeda, D. Umeno, K. Saito, F. Koide, E. Miyata, T. Sugo, Removal of boron using nylon-based chelating fibers. Ind. Eng. Chem. Res. **50**, 5727–5732 (2011)

15. K. Sekiguchi, K. Serizawa, S. Konishi, K. Saito, S. Furusaki, T. Sugo, Uranium uptake during permeation of seawater through amidoxime-group-immobilized micropores. React. Polym. **23**, 141–145 (1994)

16. E.M. Thurman, M.S. Mills. Solid-Phase Extraction (John Wiley & Sons, Inc., 1998)

17. S. Domon, S. Asai, K. Saito, K. Watanabe, T. Sugo, Selection of the alkylamino group introduced into the polymer chain grafted onto a porous membrane for the impregnation of an acidic extractant. J. Membr. Sci. **262**, 153–158 (2005)

18. S. Asai, K. Watanabe, T. Sugo, K. Saito, Interaction between an acidic extractant and an octadecylamino group introduced into a grafted polymer chain. Sep. Sci. Technol. **40**, 3349–3364 (2005)

19. S. Asai, K. Watanabe, T. Sugo, K. Saito, Preparation of an extractant-impregnated porous membrane for the high-speed separation of a metal ion. J. Chromatogr. A **1094**, 158–164 (2005)

20. K. Sawaki, S. Domon, S. Asai, K. Watanabe, T. Sugo, K. Saito, Impregnation of an acidic extractant cyanex 272 to the alkylamino group and alkylthiol group introduced into the polymer chain grafted onto a porous membrane. Membrane (Maku) **32**, 109–115 (2007)

21. S. Asai, K. Watanabe, K. Saito, T. Sugo, Preparation of Aliquat 336-impregnated porous membrane. J. Membr. Sci. **281**, 195–202 (2006)

22. S. Asai, K. Watanabe, T. Sugo, K. Saito, Effects of Aliquant 336 concentration and solvent composition on amount of Aliquat 336 impregnated and liquid permeability of Aliquat 336-impregnated porous hollow-fiber membrane. Membrane (Maku) **32**, 168–174 (2007)

23. K. Sawaki, S. Asai, K. Watanabe, T. Sugo, K. Saito, Impregnation of a neutral extractant to hydrophobic group introduced into the polymer chain grafted onto a porous membrane. Membrane (Maku) **33**, 32–38 (2008)

24. M. Kim, K. Saito, S. Furusaki, T. Sugo, J. Okamoto, Water flux and protein adsorption of a hollow fiber modified with hydroxyl groups. J. Membr. Sci. **56**, 289–302 (1991)

25. M. Kim, K. Saito, S. Furusaki, T. Sugo, Comparison of BSA adsorption and Fe sorption to the diol group and tannin immobilized onto a microfiltration membrane. J. Membr. Sci. **85**, 21–28 (1993)

26. M. Kim, J. Kojima, K. Saito, S. Furusaki, T. Sugo, Reduction of nonselective adsorption of proteins by hydrophilization of microfiltration membranes by radiation-induced grafting. Biotechnol. Prog. **10**, 114–120 (1994)

27. S. Matsuno, A. Iwanade, D. Umeno, K. Saito, H. Ito, M. Sakamoto, Carboxybetaine-group immobilized onto pore surface reduced protein adsorption to porous membrane. Membrane (Maku) **35**, 86–92 (2010)

28. K. Ishihara, T. Ueda, N. Nakabayashi, Preparation of phospholipid polymers and their properties as polymer hydrogel membranes. Polym. J. **22**, 355–360 (1990)

29. S. Tsuneda, K. Saito, S. Furusaki, T. Sugo, High-throughput processing of protein using a porous and tentacle anion-exchange membrane. J. Chromatogr. A **689**, 211–218 (1995)

30. S. Tsuneda, H. Kagawa, K. Saito, T. Sugo, Hydrodynamic evaluation of three-dimensional adsorption of protein to a polymer brush grafted onto a porous substrate. J. Colloid Interface Sci. **176**, 95–100 (1995)

31. S. Tsuneda, K. Saito, T. Sugo, K. Makuuchi, Protein adsorption characteristics of porous and tentacle anion-exchange membrane prepared by radiation-induced graft polymerization. Radiat. Phys. Chem. **46**, 239–245 (1995)

32. S. Matoba, S. Tsuneda, K. Saito, T. Sugo, Highly efficient enzyme recovery using a porous membrane with immobilized tentacle polymer chains. Nat. Biotechnol. **13**, 795–797 (1995)
33. N. Kubota, S. Miura, K. Saito, K. Sugita, K. Watanabe, T. Sugo, Comparison of protein adsorption by anion-exchange interaction onto porous hollow-fiber membrane and gel bead-packed bed. J. Membr. Sci. **117**, 135–142 (1996)
34. N. Kubota, Y. Konno, S. Miura, K. Saito, K. Sugita, K. Watanabe, T. Sugo, Comparison of two convection-aided protein adsorption methods using porous membranes and perfusion beads. Biotechnol. Prog. **12**, 869–872 (1996)
35. N. Kubota, Y. Konno, K. Saito, K. Sugita, K. Watanabe, T. Sugo, Module performance of anion-exchange porous hollow-fiber membranes for high-speed protein recovery. J. Chromatogr. A **782**, 159–165 (1997)
36. N. Kubota, Y. Konno, K. Saito, K. Sugita, K. Watanabe, T. Sugo, Protein adsorption and elution performances of modules consisting of porous anion-exchange hollow-fiber membranes. Membrane (Maku) **22**, 105–110 (1997)
37. N. Sasagawa, K. Saito, K. Sugita, T. Ogasawara, T. Sugo, Adsorption characteristics of binary proteins onto anion-exchange porous hollow-fiber membrane. J. Ion Exchange **9**, 74–80 (1998)
38. I. Koguma, K. Sugita, K. Saito, T. Sugo, Multilayer binding of proteins to polymer chains grafted onto porous hollow-fiber membranes containing different anion-exchange groups. Biotechnol. Prog. **16**, 456–461 (2000)
39. K. Hagiwara, S. Yonedu, K. Saito, T. Shiraishi, T. Sugo, T. Tojyo, E. Katayama, High-performance purification of gelsolin from plasma using anion-exchange porous hollow-fiber membrane. J. Chromatogr. B **821**, 153–158 (2005)
40. S. Yonedu, K. Saito, E. Katayama, T. Tojyo, T. Shiraishi, T. Sugo, Affinity elution of gelsolin adsorbed onto an anion-exchange porous membrane. Membrane (Maku) **30**, 269–274 (2005)
41. T. Yoshikawa, K. Hagiwara, K. Saito, E. Katayama, T. Tojyo, T. Sugo, Comparison of gelsolin purification performance between anion-exchange-graft-chain-containing porous membrane and anion-exchange bead-packed bed. J. Ion Exchange **18**, 2–8 (2007)
42. A. Nide, S. Kawai-Noma, D. Umeno, K. Saito, Reduction of buffer volume used in regeneration of anion-exchange porous hollow-fiber membrane by site-controlled introduction of anion-exchange group into graft chain. Membrane (Maku) **39**, 258–263 (2014)
43. K. Kobayashi, S. Tsuneda, K. Saito, H. Yamagishi, S. Furusaki, T. Sugo, Preparation of microfiltration membranes containing anion-exchange groups. J. Membr. Sci. **76**, 209–218 (1993)
44. H. Shinano, S. Tsuneda, K. Saito, S. Furusaki, T. Sugo, Ion exchange of lysozyme during permeation across a microporous sulfopropyl-group-containing hollow fiber. Biotechnol. Prog. **9**, 193–198 (1993)
45. S. Tsuneda, H. Shinano, K. Saito, S. Furusaki, T. Sugo, Binding of lysozyme onto a cation-exchange microporous membrane containing tentacle-type grafted polymer branches. Biotechnol. Prog. **10**, 76–81 (1994)
46. N. Sasagawa, K. Saito, K. Sugita, S. Kunori, T. Sugo, Ionic crosslinking of SO_3H-group-containing graft chains helps to capture lysozyme in a permeation mode. J. Chromatogr. A **848**, 161–168 (1999)
47. D. Okamura, K. Saito, K. Sugita, M. Tamada, T. Sugo, Solvent effect on protein binding by polymer brush grafted onto porous membrane. J. Chromatogr. A **953**, 101–109 (2002)
48. D. Okamura, K. Saito, K. Sugita, M. Tamada, T. Sugo, Effect of alcohol solvent for glycidyl methacrylate in radiation-induced graft polymerization on performance of cation-exchange porous membranes. Membrane (Maku) **27**, 196–201 (2002)
49. A. Iwanade, D. Umeno, K. Saito, T. Sugo, Protein binding to amphoteric polymer brushes grafted onto a porous hollow-fiber membrane. Biotechnol. Prog. **23**, 1425–1430 (2007)
50. A. Iwanade, T. Nomoto, D. Umeno, K. Saito, T. Sugo, Protein binding characteristics of amphoteric polymer brushes grafted onto porous hollow-fiber membrane. J. Ion Exchange **18**, 492–497 (2007)

51. A. Iwanade, D. Umeno, K. Saito, T. Sugo, Dependence of protein binding capacity of dimethylamino-γ-butyric-acid (DMGABA)-immobilized porous membrane on composition of solvent used for DMGABA immobilization. Radiat. Phys. Chem. **87**, 53–58 (2013)

52. T. Kawai, K. Sugita, K. Saito, T. Sugo, Extension and shrinkage of polymer brush grafted onto porous membrane induced by protein binding. Macromolecules **33**, 1306–1309 (2000)

53. K. Miyoshi, K. Saito, T. Shiraishi, T. Sugo, Introduction of taurine into polymer brush grafted onto porous hollow-fiber membrane. J. Membr. Sci. **264**, 97–103 (2005)

54. N. Kubota, M. Kounosu, K. Saito, K. Sugita, K. Watanabe, T. Sugo, Preparation of a hydrophobic porous membrane containing phenyl groups and its protein adsorption performance. J. Chromatogr. A **718**, 27–34 (1995)

55. N. Kubota, M. Kounosu, K. Saito, K. Sugita, K. Watanabe, T. Sugo, Control of phenyl-group site introduced on the graft chain for hydrophobic interaction chromatography. React. Polym. **29**, 115–122 (1996)

56. N. Kubota, M. Kounosu, K. Saito, K. Sugita, K. Watanabe, T. Sugo, Protein adsorption and elution performances of porous hollow-fiber membranes containing various hydrophobic ligands. Biotechnol. Prog. **13**, 89–95 (1997)

57. N. Kubota, M. Kounosu, K. Saito, K. Sugita, K. Watanabe, T. Sugo, Repeated use of a hydrophobic ligand-containing porous membrane for protein recovery. J. Membr. Sci. **134**, 67–73 (1997)

58. S. Nishiyama, A. Goto, K. Saito, K. Sugita, M. Tamada, T. Sugo, T. Funami, Y. Goda, S. Fujimoto, Concentration of 17β-estradiol using an immunoaffinity porous hollow-fiber membrane. Anal. Chem. **74**, 4933–4936 (2002)

59. M. Kim, S. Kiyahara, S. Konishi, S. Tsuneda, K. Saito, T. Sugo, Ring-opening reaction of poly-GMA chain grafted onto a porous membrane. J. Membr. Sci. **117**, 33–38 (1996)

60. S. Kiyohara, M. Kim, Y. Toida, K. Saito, K. Sugita, T. Sugo, Selection of a precusor monomer for the introduction of affinity ligands onto a porous membrane by radiation-induced graft polymerization. J. Chromatogr. A **758**, 209–215 (1997)

61. H. Iwata, K. Saito, S. Furusaki, T. Sugo, J. Okamoto, Adsorption characteristics of an immobilized metal affinity membrane. Biotechnol. Prog. **7**, 412–418 (1991)

62. K. Kin, K. Hagiwara, D. Umeno, K. Saito, T. Sugo, Purification of His-tagged protein using an immobilized nickel affinity porous hollow-fiber membrane. Membrane (Maku) **34**, 233–238 (2009)

63. Y. Monma, D. Umeno, K. Saito, T. Sugo, Binding of phosphotyrosine to gallium-ion-immobilized porous hollow-fiber membrane. Membrane (Maku) **35**, 242–247 (2010)

64. M. Kim, K. Saito, S. Furusaki, T. Sugo, I. Ishigaki, Adsorption and elution of bovine gamma-globulin using an affinity membrane containing hydrophobic amino acids as ligands. J. Chromatogr. **585**, 45–51 (1991)

65. M. Kim, K. Saito, S. Furusaki, T. Sugo, I. Ishigaki, Protein adsorption capacity of a porous phenylalanine-containing membrane based on a polyethylene matrix. J. Chromatogr. **586**, 27–33 (1991)

66. S. Matsuno, D. Umeno, M. Miyazaki, Y. Suzuta, K. Saito, T. Yamashita, Immobilization of an esterase inhibitor on a porous hollow-fiber membrane by radiation-induced graft polymerization for developing a diagnostic tool for feline kidney diseases. Biosci. Biotechnol. Biochem. **77**, 2061–2064 (2013)

67. S. Kiyohara, M. Sasaki, K. Saito, K. Sugita, T. Sugo, Radiation-induced grafting of phenylalanine-containing monomer onto a porous membrane. React. Polym. **31**, 103–110 (1996)

68. S. Kiyohara, M. Sasaki, K. Saito, K. Sugita, T. Sugo, Amino acid addition to epoxy-group-containing polymer chain grafted onto a porous membrane. J. Membr. Sci. **109**, 87–92 (1996)

69. A. Shibasaki, Y. Irimoto, M. Kim, K. Saito, K. Sugita, T. Baba, I. Honjyo, S. Moriyama, T. Sugo, Selective binding of docosahexaenoic acid ethyl ester to a silver-ion-loaded porous hollow-fiber membrane. JAOCS **76**, 771–775 (1999)

70. I. Ozawa, M. Kim, K. Saito, K. Sugita, T. Baba, S. Moriyama, T. Sugo, Purification of docosahexaenoic acid ethyl ester using silver-ion immobilized porous hollow-fiber membrane module. Biotechnol. Prog. **17**, 893–896 (2001)
71. S. Konishi, K. Saito, S. Furusaki, T. Sugo, Binary metal-ion sorption during permeation through chelating porous membrane. J. Membr. Sci. **111**, 1–6 (1996)
72. J.S. Chaberek, A.E. Martell, Stability of metal chelates. I. Iminodiacetic and iminodipropinic acids. J. Am. Chem. Soc. **74**, 5052–5056 (1952)
73. H. Kawakita, K. Sugita, K. Saito, M. Tamada, T. Sugo, H. Kawamoto, Optimization of reaction conditions in production of cycloisomaltooligosaccharides using enzyme immobilized in multilayers onto pore surface of porous hollow-fiber membranes. J. Membr. Sci. **205**, 175–182 (2002)
74. T. Kawai, H. Kawakita, K. Sugita, K. Saito, M. Tamada, T. Sugo, H. Kawamoto, Conversion of dextran to cycloisomaltooligosaccharides using enzyme-immobilized porous hollow-fiber membrane. J. Agric. Food Sci. **50**, 1073–1076 (2002)
75. H. Kawakita, K. Sugita, K. Saito, M. Tamada, T. Sugo, H. Kawamoto, Production of cycloisomaltooligosaccharides from dextran using enzyme immobilized in multilayers onto porous membranes. Biotechnol. Prog. **18**, 465–469 (2002)
76. H. Kawakita, K. Saito, K. Sugita, M. Tamada, T. Sugo, H. Kawamoto, Skin-layer formation of porous membrane by immobilized dextransucrase. AIChE J. **50**, 696–700 (2004)
77. S. Kobayashi, S. Yonedu, H. Kawakita, K. Saito, K. Sugita, M. Tamada, T. Sugo, W. Lee, Highly multilayered urease decomposes highly concentrated urea. Biotechnol. Prog. **19**, 396–399 (2003)
78. M. Nakamura, K. Saito, K. Sugita, T. Sugo, Application of crosslinked-aminoacylase-multilayered membranes to bioreactor. Membrane (Maku) **23**, 316–321 (1998)
79. T. Kawai, M. Nakamura, K. Sugita, K. Saito, T. Sugo, High conversion in asymmetric hydrolysis during permeation through enzyme-multilayered porous hollow-fiber membranes. Biotechnol. Prog. **17**, 872–875 (2001)
80. T. Kawai, K. Saito, K. Sugita, T. Sugo, H. Misaki, Immobilization of ascorbic acid oxidase in multilayers onto porous hollow-fiber membrane. J. Membr. Sci. **191**, 207–213 (2001)
81. S. Miura, N. Kubota, H. Kawakita, K. Saito, K. Sugita, K. Watanabe, T. Sugo, High-throughput of hydrolysis of starch during permeation across amylase-immobilized porous hollow-fiber membrane. Radiat. Phys. Chem. **63**, 143–149 (2002)
82. A. Fujita, H. Kawakita, K. Saito, K. Sugita, M. Tamada, T. Sugo, Production of tripeptide from gelatin using collagenase-immobilized porous hollow-fiber membrane. Biotechnol. Prog. **19**, 1365–1367 (2003)
83. I. Koguma, M. Nakamura, K. Saito, K. Sugita, S. Kiyohara, T. Sugo, Chiral separation of DL-tryptophan using bovine-serum-albumin-multilayered porous hollow-fiber membrane. Kagaku Kogaku Ronbunsyu **24**, 458–461 (1998)
84. M. Nakamura, S. Kiyohara, K. Saito, K. Sugita, T. Sugo, Chiral separation of DL-tryptophan using porous membranes containing multilayered bovine serum albumin crosslinked with glutaraldehyde. J. Chromatogr. A **822**, 53–58 (1998)
85. S. Kiyohara, M. Nakamura, K. Saito, K. Sugita, T. Sugo, Binding of DL-tryptophan to BSA adsorbed in multilayers by polymer chains grafted onto a porous hollow-fiber membrane in a permeation mode. J. Membr. Sci. **152**, 143–149 (1999)
86. M. Nakamura, S. Kiyohara, K. Saito, K. Sugita, T. Sugo, High resolution of DL-tryptophan at high flow rates using a bovine serum albumin-multilayered porous hollow-fiber membrane. Anal. Chem. **71**, 1323–1325 (1999)
87. H. Ito, M. Nakamura, K. Saito, K. Sugita, T. Sugo, Comparison of L-tryptophan binding capacity of BSA captured by a polymer brush with that of BSA adsorbed onto a gel network. J. Chromatogr. A **925**, 41–47 (2001)
88. K. Saito, K. Saito, K. Sugita, M. Tamada, T. Sugo, Convection-aided collection of metal ions using chelating porous flat-sheet membranes. J. Chromatogr. A **954**, 277–283 (2002)

89. K. Yamashiro, K. Miyoshi, R. Ishihara, D. Umeno, K. Saito, T. Sugo, S. Yamada, H. Fukunaga, M. Nagai, High-throughput solid-phase extraction of metal ions using an iminodiacetate chelating porous disk prepared by graft polymerization. J. Chromatogr. A **1176**, 37–42 (2007)

90. G. Wada, R. Ishihara, K. Miyoshi, D. Umeno, K. Saito, S. Asai, S. Yamada, H. Hirota, Effect of chelating group density of crosslinked graft chain on dynamic binding capacity for metal ions. J. Ion Exchange **22**, 47–52 (2011)

91. G. Wada, R. Ishihara, K. Miyoshi, D. Umeno, K. Saito, S. Asai, S. Yamada, H. Hirota, Crosslinked-chelating porous sheet with high dynamic binding capacity of metal ions. Solv. Extr. Ion Exch. **31**, 210–220 (2013)

92. K. Yamashiro, K. Miyoshi, R. Ishihara, K. Yasuno, D. Umeno, K. Saito, T. Sugo, S. Yamada, M. Sugiura, H. Fukunaga, M. Nagai, Protein purification using immobilized metal affinity porous sheet. J. Ion Exchange **19**, 101–106 (2008)

93. S. Asai, M. Magara, S. Sakurai, N. Shinohara, K. Saito, T. Sugo, Rapid separation of actinides using an anion-exchange polymer chain grafted onto a porous sheet. J. Ion Exchange **18**, 486–491 (2007)

94. R. Ishihara, D. Umeno, K. Saito, S. Asai, S. Sakurai, N. Shinohara, T. Sugo, Preparation of extractant-impregnated porous sheets for high-speed separation of radionuclides. J. Ion Exchange **18**, 480–485 (2007)

95. S. Asai, M. Magara, N. Shinohara, S. Yamada, M. Nagai, K. Miyoshi, K. Saito, Separation of U and Pu in spent nuclear fuel sample using anion-exchange-group-introduced porous polymer sheet for ICP-MS determination. Talanta **77**, 695–700 (2008)

96. S. Asai, T. Kimura, K. Miyoshi, K. Saito, S. Yamada, H. Hirota, Application of diethylamino-group-containing porous-polymeric-disk-packed cartridge to separation of U in urine sample. J. Ion Exchange **21**, 334–339 (2010)

97. S. Asai, Y. Hanzawa, M. Konda, D. Suzuki, M. Magara, T. Kimura, R. Ishihara, K. Saito, S. Yamada, H. Hirota, Preparation of microvolume anion-exchange cartridge for inductively coupled plasma mass spectrometry-based determination of ^{237}Np content in spent nuclear fuel. Anal. Chem. **88**, 3149–3155 (2016)

98. R. Ishihara, S. Asai, S. Otosaka, S. Yamada, H. Hirota, K. Miyoshi, D. Umeno, K. Saito, Dependence of lanthanide-ion binding performance on HDEHP concentration in HDEHP impregnation to porous sheet. Solv. Extr. Ion Exch. **30**, 171–180 (2012)

99. R. Tanaka, R. Ishihara, K. Miyoshi, D. Umeno, K. Saito, S. Asai, S. Yamada, H. Hirota, Modification of a hydrophobic-ligand-containing porous sheet using tri-n-octylphosphine oxide, and its adsorption/elution of bismuth ions. React. Funct. Polym. **70**, 986–990 (2010)

100. J. Zhang, Y. Nozaki, Behavior of rare earth elements in seawater at the ocean margin: study along the slopes of the Sagami and Nankai troughs near Japan. Geochim. Cosmochim. Acta **62**, 1307–1317 (1998)

Chapter 4
Revolution in the Form of Polymeric Adsorbents 2: Fibers, Films, and Particles

Abstract Commercially available adsorbents are in the form of beads, granules, and short-length fibers. A 15-cm-diameter bobbin consisting of a fiber can be modified by radiation-induced graft polymerization. The fiber of the resultant bobbin can be fabricated into a wound filter, a nonwoven fabric, or a braid depending on the conditions of practical separation. In this chapter, examples of the application of functional fibers are the recovery of uranium from seawater using a chelating-group-immobilized fiber and the resolution of neodymium and dysprosium using an extractant-impregnated fiber. In addition, a polyethylene film is modified into ion-exchange membranes installed in an electrodialyzer for the production of edible salt.

Keywords Wound filter · Nonwoven fabrics · Recovery of uranium from seawater · Resolution of neodymium and dysprosium

4.1 Fibers

A fiber with a small diameter is regarded as a strand of small beads and is easy to handle without being packed into a bed. Polymeric adsorbents in fiber form, such as 50-μm-diameter ion-exchange fibers, have advantages over those in bead form in that the diffusional mass-transfer path of ions is shorter and the specific surface area of the adsorbent is larger. Additionally, a polymeric fiber can be molded into various products, such as braids and wound filters. Braids consisting of cobalt-ferrocyanide-impregnated fibers have been installed at various sites in the TEPCO's Fukushima Daiichi Nuclear Power Plant (NPP) for the removal of radioactive cesium ions from contaminated streams.

Radiation-induced graft polymerization can be used to prepare polymeric fibrous adsorbents [1]. Thus far, strongly acidic cation-exchange and strongly basic anion-exchange fibers have been easily prepared by graft polymerization of sodium styrene sulfonate (SSS) and vinyl benzyl trimethylammonium chloride (VBTAC), respectively, onto electron-beam- or gamma ray-irradiated nylon-6 fiber, as shown

© Springer Nature Singapore Pte Ltd. 2018
K. Saito et al., *Innovative Polymeric Adsorbents*,
https://doi.org/10.1007/978-981-10-8563-5_4

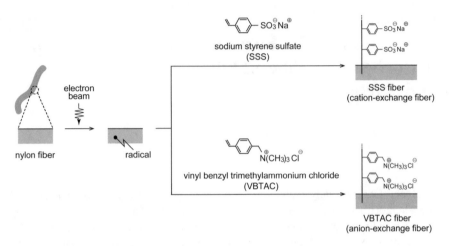

Fig. 4.1 Preparation schemes for ion-exchange fibers

in Fig. 4.1. In these processes, the use of an aqueous solution of SSS or VBTAC as a vinyl monomer solution is favorable for mass production of ion-exchange fibers. A higher degree of grafting of vinyl monomers that originally contain functional groups leads to a higher density of functional groups, where the degree of grafting is adjustable by altering radiation dose, vinyl monomer concentration, grafting temperature, and time.

4.1.1 Removal of Radioactive Substances from Contaminated Water at TEPCO's Fukushima Daiichi NPP

The East Japan Earthquake with a magnitude of 9.0 and the tsunami that followed on 11 March 2011 caused the meltdown of three reactors of the TEPCO's Fukushima Daiichi NPP. Radionuclides such as iodine-131, cesium-137, and strontium-90 produced by the fission of nuclear fuels were emitted both on-site and off-site. Since then, water contaminated with the radionuclides has mainly been stored at a rate of 150–300 m^3 per day in tanks with 1000 m^3 capacity installed at the NPP. The seawater in front of the seawater-intake area of Units 1–4 at the NPP has been contaminated with radioactive ions such as ionic Cs-137 and Sr-90 that leaked from the reactors. Nonradioactive cesium and strontium also dissolved in seawater at respective concentrations of 0.0003 and 8 mg/L.

Adsorption is a feasible method of removing radioactive species from water contaminated with radionuclides. Because an adsorbent cannot distinguish radioactive ions from nonradioactive ions, high adsorption capacity and the capability to collect both ionic types are crucial requisites for adsorbents. In addition, fibrous adsorbents are desirable for the treatment of contaminated water because undefined volumes of water such as groundwater and drainage are also treated at the NPP.

Insoluble metal ferrocyanides have been reported to specifically capture cesium ions through an ion-exchange interaction [2–5]; the metals include cobalt, nickel, and iron. Insoluble metal ferrocyanides form microparticles with an average size of 50 nm; therefore, for practical cesium removal, insoluble metal ferrocyanides have been impregnated onto various supports such as ion-exchange resin beads [6–9], silica gel [10, 11], and zeolites [12]. Thus far, these adsorbents in bead or granule form have been packed into beds, through which radionuclide-containing solutions flow. However, fibrous adsorbents are versatile and convenient for easy operation at sites for processing contaminated water. To remove radioactive cesium ions from the seawater of the seawater-intake area, we proposed a novel adsorptive braid consisting of cobalt-ferrocyanide-impregnated fibers [13–21]. These fibers with small diameters enhance the mass transfer of cesium ions from the bulk of liquid to cobalt-ferrocyanide particles, resulting in rapid removal of cesium ions from seawater.

The process of impregnating cobalt ferrocyanide into nylon fibers consists of the following four steps [18] (Fig. 4.2). (1) Grafting of glycidyl methacrylate: A nylon fiber, previously irradiated with gamma rays at a dose of 20 kGy, is immersed in 15% (v/v) glycidyl methacrylate (GMA)/methanol at 333 K. The degree of GMA grafting is set at 110%. (2) Addition of triethylene diamine (TEDA) to epoxy groups of GMA-grafted fibers: A GMA-grafted fiber is immersed in 1 M TEDA solution at 333 K for 5 h. The density of immobilized TEDA is set at 1.7 mmol/g. (3) Binding of ferrocyanide ions, $[Fe(CN)_6]^4$: A TEDA-immobilized fiber, which was previously conditioned by immersion in 0.1 M HCl for 30 min, is immersed in 0.002–0.031 M potassium ferrocyanide solution for 1 h. (4) Precipitation of cobalt ferrocyanide: After a ferrocyanide-ion-bound fiber is reacted with 0.25 M $CoCl_2$ solution at 303 K for 10 h to precipitate cobalt ferrocyanide into the grafted polymer chain, the resultant fiber is washed repeatedly with pure water and dried in vacuum. The impregnation percentage was evaluated from the mass gain of the fiber to be in the range from 2 to 11%. Profiles of elemental Co and Fe across a cross section of the cobalt-ferrocyanide-impregnated fiber determined using a scanning electron microscopy-energy dispersive X-ray spectroscopy (SEM-EDX) system demonstrated that the elemental Co and Fe were concentrated in the periphery of the fiber with the graft chains, but not in the core consisting of the nylon matrix and the graft chains [20].

The essential selection criteria for the adsorbent for the removal of cesium ions are adsorption rate and capacity. In most cases, the adsorbents are repeatedly used through a cycle of adsorption, elution, and regeneration; however, once radioactive cesium ions are adsorbed onto the adsorbent, the adsorbent is reduced in volume and stored in a high-integrity container (HIC) without the elution of cesium ions. The removal rate of cesium ions from seawater in a batch mode was compared between a cobalt-ferrocyanide-impregnated fiber and a zeolite, chabazite Ionsiv

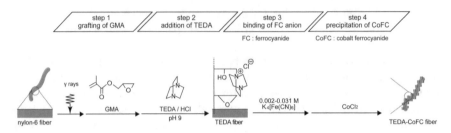

Fig. 4.2 Preparation scheme for insoluble cobalt-ferrocyanide-impregnated fiber for the removal of cesium ions

IE-96, in which an initial cesium concentration and a liquid-to-fiber ratio were set at 10 mg-Cs/L or 10 ppm and 100, respectively. These experimental conditions were adopted by the Japan Atomic Energy Society as recommended immediately after the accident. As shown in Fig. 4.3, the cobalt-ferrocyanide-impregnated fiber exhibited a rapid decrease in the concentration of cesium ions in seawater within 30 min to below the detection limit of ICP-MS, i.e., 0.2 mg-Cs/L, whereas the zeolite showed a slow decrease in the concentration to 1 ppm after 12 h.

Adsorption isotherms of the adsorbents for cesium ions in seawater were also compared between the cobalt-ferrocyanide-impregnated fiber and the zeolite. As shown in Fig. 4.4, the cobalt-ferrocyanide-impregnated fiber exhibited a more favorable adsorption equilibrium than the zeolite: A larger amount of cesium ions was adsorbed at a lower concentration of cesium ions in seawater. Cesium at an extremely low concentration, e.g., 2.8×10^{-8} mg/L, equivalent to 90 Bq/L of the regulatory concentration of cesium, dissolves in seawater that contains stable cesium ions at a concentration of 0.0003 mg/L. In principle, a "jungle-gym"-like lattice structure of cobalt ferrocyanide captures cesium ions more specifically than the microporous structure of the zeolite.

Fig. 4.3 Comparison of cesium removal rate between adsorptive fiber and zeolite

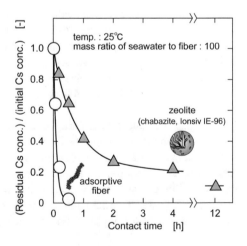

Fig. 4.4 Comparison of adsorption isotherm between adsorptive fiber and zeolite in seawater

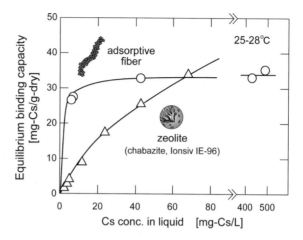

To better understand the structure of cobalt ferrocyanide impregnated into the fiber, after the initial impregnation of cobalt ferrocyanide into the fiber, ferrocyanide ions were rebound to the TEDA moieties of the cobalt-ferrocyanide-impregnated fiber [21]. The amount of ferrocyanide ions rebound to the fiber decreased as the amount of cobalt ferrocyanide initially impregnated was increased. Furthermore, the difference between the amount of ferrocyanide ions initially bound and that of ferrocyanide ions rebound was directly proportional to the amount of cobalt ferrocyanide initially impregnated onto the fiber; the plot of these results is a straight line with a slope of unity (Fig. 4.5). This finding suggests that all of the TEDA moieties capturing ferrocyanide ions of the graft chain are involved in the nucleation and growth of cobalt ferrocyanide onto the graft chain: The TEDA moieties of the graft chain are expected to be occupied by the microparticles of cobalt ferrocyanide. A possible structure representing the impregnation of cobalt ferrocyanide onto graft chains is illustrated in Fig. 2.27 along with a SEM image of the fiber surface. The microparticles may become entangled and penetrate within the graft chains as a result of multipoint electrostatic interactions between the positive charge of TEDA moieties and the negative surface charge of cobalt-ferrocyanide microparticles.

A bobbin of cobalt-ferrocyanide-impregnated fiber with a mass of 100 kg was produced, from which a number of adsorptive braids were formed [15]. On June 11, 2013, braids with a diameter of approximately 8 cm and a total length of approximately 70 m were immersed in the seawater-intake pit of Unit 3 of the NPP. In addition, from December 2014, braids arranged as a curtain were immersed into the seawater-intake area to confirm their performance. In addition, these braids have been used at 20 sites in the drainage system of the NPP.

Radioactive strontium must be removed from seawater contaminated with radionuclides with an extremely high efficiency for two reasons: (1) Nonradioactive or stable strontium originally dissolves in seawater at a concentration of approximately 8 mg-Sr/L, and (2) calcium and magnesium ions that belong to the same alkaline-earth family as strontium ions dissolve in seawater at 600- and 100-fold higher molar concentrations, respectively, than strontium ions.

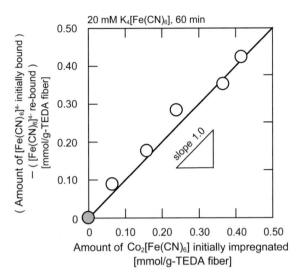

Fig. 4.5 Difference between the amount of ferrocyanide ion initially bound and that of ferrocyanide ions rebound versus the amount of cobalt ferrocyanide initially impregnated into the fiber. Reprinted with permission from Ref. [21]. Copyright 2016 Taylor & Francis Ltd.

Sodium titanate is a candidate as an adsorbent capable of removing strontium ions from seawater [22]. Sodium titanate was impregnated into the nylon fiber by radiation-induced graft polymerization and subsequent chemical modifications [23–29] (Fig. 4.6). First, dimethylaminopropyl acrylamide as an originally anion-exchange-group-containing vinyl monomer was graft-polymerized onto an electron-beam-irradiated nylon fiber, followed by binding of peroxotitanium complex anions to the anion-exchange groups of the graft chains. Then, bound titanium species were converted into insoluble sodium titanate by reaction with sodium hydroxide [27–29].

Water contaminated with radionuclides has been generated by the contact between the water used to cool the meltdown fuel and groundwater since the accident at the NPP on March 11, 2011. Sixty-two different radioactive substances have been removed by using advanced liquid processing system (ALPS), which consists of a series of many adsorption columns, as shown in Fig. 4.7. Water from which substances have been removed to below the maximum allowable radioactivity level is being stored in tanks at the NPP.

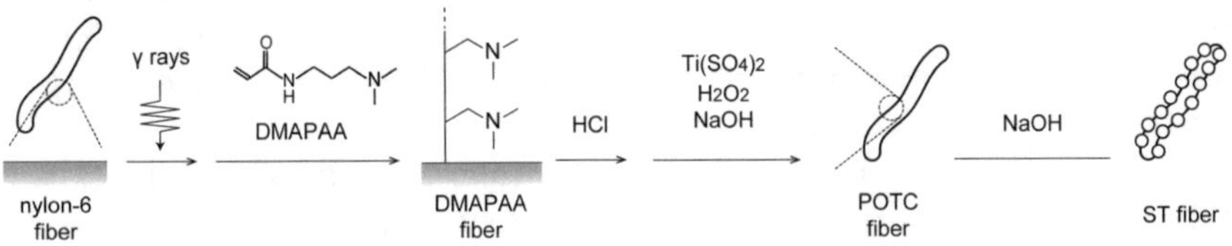

Fig. 4.6 Preparation scheme for sodium titanate-impregnated fiber for the removal of strontium ions

Radioactive ruthenium (Ru) is difficult to remove from the liquid stream by ALPS, so it is removed at the final stage of ALPS using adsorption columns (Fig. 4.7). Ru is one of the platinum group metals (PGMs). We have succeeded in collecting palladium (Pd) ions using an adenine- or guanine-immobilized porous hollow-fiber membrane [30]; therefore, nucleic-acid-base-immobilized adsorbents are promising for the removal of Ru ions [31].

The immobilization scheme for adenine onto nylon-6 fiber is shown in Fig. 4.8. The resultant fiber with a density of immobilized adenine of 1.2 mmol/g was immersed in 10 mg-Ru/L $RuCl_3$ solution (pH 2.0) at a liquid-to-fiber ratio of 100 mL/g. The decay in the Ru concentration was determined at various NaCl concentrations of up to 0.5 mol/L and temperatures from 8 to 60 °C. The concentration decay curves are shown in Fig. 4.9. As the NaCl concentration and temperature increased, the decrease in Ru concentration accelerated. This is because Ru-chloro complexes more readily react with the adenine moiety immobilized by the graft chain than Ru-aquo complexes at higher temperatures.

The decay curves were kinetically analyzed using a pseudosecond-order reaction model to evaluate the reaction rate constant. The reaction rate constants at various temperatures and concentrations are plotted in Fig. 4.10 in accordance with the Arrhenius equation. The activation energy calculated from the slope of the resultant

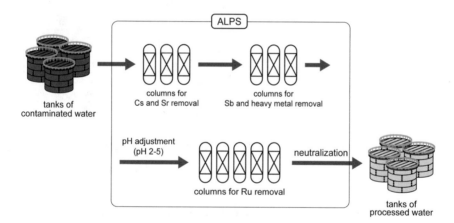

Fig. 4.7 Schematic illustration of ALPS

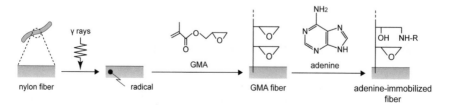

Fig. 4.8 Immobilization scheme for adenine onto nylon-6 fiber

Fig. 4.9 Concentration decay of Ru ionic species in NaCl solution in batch mode

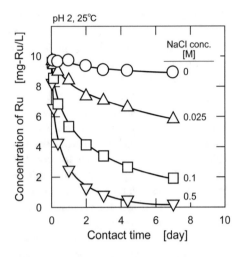

Fig. 4.10 Arrhenius plot of pseudosecond-order reaction rate constant

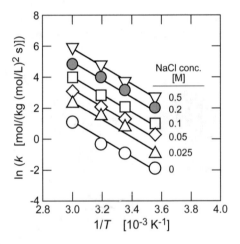

straight line was 45 kJ/mol; this indicates that the overall adsorption of Ru onto the adenine-immobilized fiber is governed by the intrinsic complexation of Ru ionic species with the immobilized adenine moiety as well as by diffusional mass transfer.

4.1.2   Removal of Urea from Ultrapure Water

A large amount of ultrapure water is consumed at integrated circuit (IC) manu-facturing factories in a wet process to thoroughly wash IC surfaces. To produce ultrapure water, the factories are commonly located along a river. The production of ultrapure water requires many unit operations: aggregation, reverse osmosis (RO)-membrane filtration, electrical desalination, UV irradiation, ion exchange, and ultrafiltration (UF)-membrane filtration (Fig. 4.11). Of all the ions and molecules dissolved in river streams, urea derived from fertilizers or livestock excretion is quite difficult to remove because the urea molecule is small and electrically neutral.

Indirect removal of urea with urease is used because an adsorbent capable of directly binding urea has not yet been developed. Using a urease-immobilized material, urea is decomposed into ammonium ions (NH_4^+) and carbon dioxide (CO_2). Subsequently, ammonium ions are captured by cation-exchange resins, and carbon dioxide is stripped under reduced pressure.

To attain a high-throughput removal of urea dissolved in ultrapure water, the replacement of urease-immobilized beads with urease-immobilized fibers is desirable. The preparation of urease-immobilized fibers consists of four steps [32], as shown in Fig. 4.12. First, nylon fibers are irradiated with an electron beam. Second, an anion-exchange group-containing vinyl monomer, dimethylaminoethyl methacrylate (DMAEMA), is graft-polymerized onto the irradiated nylon fibers. Third, urease (Mr 480,000, pI 5.0) is bound to the resultant anion-exchange fibers through an electrostatic interaction. Finally, bound urease is crosslinked to other bound urease molecules using transglutaminase.

A fiber with of urease immobilized at a density of 41 mg/g was cut into about 1 cm lengths and packed into a bed with an inner diameter of 5.5 mm and a height of 30 mm. When a urea solution was passed through the fiber bed at a flow rate in

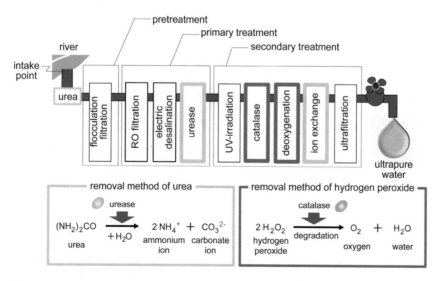

Fig. 4.11 Removal of urea and hydrogen peroxide to produce ultrapure water using urease- and catalase-immobilized fibers

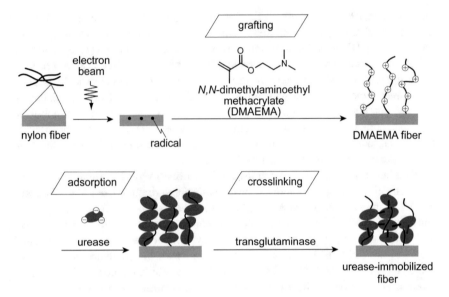

Fig. 4.12 Immobilization of urease to a polymer chain grafted onto nylon fiber

Fig. 4.13 Percentage
hydrolysis of urea as a
function of space velocity

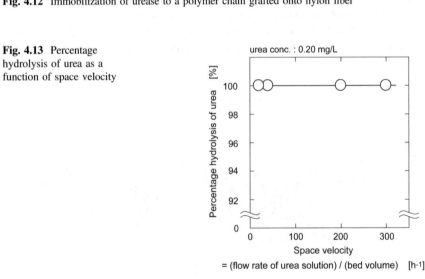

the range from 13 to 190 mL/h, i.e., in the SV range from 20 to 300 h^{-1}, the
percentage hydrolysis remained constant at 100% (Fig. 4.13): A quantitative
hydrolysis of urea in water was achieved. Catalase was also immobilized onto the
nylon fiber to hydrolyze hydrogen peroxide (H_2O_2) to produce ultrapure water [33].

4.1.3 Recovery of Uranium from Seawater

The concentration of uranium in seawater is markedly constant at 3.3 mg-U/m^3. The predominant dissolved form of uranium in seawater is a stable uranyl tricarbonate complex $UO_2(CO_3)_3^{4-}$ [34, 35]. The total uranium content of 4.5 billion tons dissolved in the world's oceans is almost 1000-fold larger than terrestrial resources of reasonable concentrations. Nuclear power plants continuously require uranium; therefore, the 4.5 billion tons of uranium in seawater will be essential for use in nuclear power facilities.

A recovery program was begun in England in the early 1960s: The recovery of uranium from seawater using an adsorption column charged with hydrous titanium oxide was suggested in 1964 by Davies et al. [36]. Seawater was pumped upward through an adsorption column. Since then, the recovery of uranium from seawater has been extensively researched to replace uranium locally deposited as terrestrial ore with uranium uniformly dissolved in seawater [37–40].

Many methods of recovery have been suggested: coprecipitation, adsorption, ion floatation, and solvent extraction. Adsorption using solid adsorbents is promising with regard to their economic and environmental impacts. Projects on uranium recovery from seawater have been carried out for two decades in Japan from the laboratory scale to an offshore plant scale; extensive development of high-performance adsorbents and feasibility studies on recovery systems led to the development of an ocean-current/wave-utilizing system using a submerged adsorption cage packed with adsorbents, which was based on the system designed by Nobukawa et al. [41–43].

The molar concentration of uranium, 1.4×10^{-5} mol/m^3, is approximately 1 part in 4×10^6 of that of magnesium, which is the representative bivalent cation in seawater. Extensive effort has been exerted to develop an adsorbent capable of separating uranium from other elements [44–52]. At present, a resin containing an amidoxime group [–C(=NOH)NH$_2$] prepared by the reaction of cyano groups with hydroxylamine (NH$_2$OH) is promising in terms of adsorption rate, capacity, durability, and production cost [53, 54].

The recovery process for uranium from seawater consists of three stages: (1) adsorption of uranium from seawater using an amidoxime resin and subsequent elution with an eluent, (2) purification of uranium from the eluent with another chelating resin, and (3) further concentration of uranium using an anion-exchange resin. Since an adsorbent should come in contact with a tremendous volume of seawater in the first step, various effective contacting systems have been suggested and evaluated. The adsorption systems for the recovery of uranium from seawater can be classified according to three factors: (1) the form of the adsorbent, i.e., spherical or fibrous; (2) the type of the adsorption bed, i.e., fixed or fluidized; and (3) the method of moving seawater, i.e., by pumping or use of ocean currents. The system consists of a combination of these three factors.

4.1.4 Amidoxime Fiber-Packed Bed Placed Along the Coast of the Pacific Ocean [55–70]

An experimental setup for uranium recovery from seawater was placed on the coast of the Pacific Ocean to acquire fundamental data at controlled temperature and flow rate of seawater [64]. We prepared amidoxime (AO) hollow fibers using radiation-induced graft polymerization. Porous polyethylene hollow fibers, provided by Mitsubishi Rayon Co. Ltd, Japan, were used as the trunk polymers for grafting. The inner and outer diameters of these hollow fibers were 0.027 and 0.038 cm, respectively, with 60% porosity. A filtration module loaded with a bundle of these hollow fibers has been commercially used as a water purification device capable of removing microorganisms and colloids (Fig. 4.14). AO hollow fibers (AO-H fibers) (Fig. 4.15) are prepared as follows [61]: radiation-induced graft polymerization of acrylonitrile onto porous hollow fibers, amidoximation, and alkaline treatment.

The amidoximation of the cyano group and subsequent alkaline treatment produce various moieties such as amidoxime, imidedioxime, and carboxyl groups, depending on the reaction time and the solvent used. Amidoxime and imidedioxime groups are reported to bind the stable uranyl tricarbonate complex in seawater as follows [48]:

Fig. 4.14 Structure of household water purifier

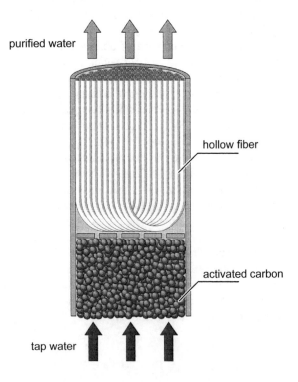

purified water

hollow fiber

activated carbon

tap water

Fig. 4.15 Amidoxime group capable of specifically collecting uranium species in seawater

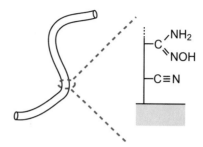

$$2HL + UO_2(CO)_3^{4-} = UO_2L_2 + 3CO_3^{2-} + 2H^+ \qquad (4.1)$$

$$H_2L + UO_2(CO)_3^{4-} = UO_2L + 3CO_3^{2-} + 2H^+ \qquad (4.2)$$

where HL and H_2L denote the amidoxime and imidedioxime groups, respectively. Alkaline treatment induces the formation of micropores favorable for diffusion of the uranyl tricarbonate complex into the AO-H fibers; this improves the overall uranium adsorption rate [63].

The continuous-flow experimental setup placed on the coast of the Pacific Ocean is shown in Fig. 4.16. Seawater was first filtered through a sand filter and then through a pleated cartridge filter having a nominal pore size of 10 μm before being pumped upward through an AO-H fiber-packed bed. The temperature of the seawater was maintained to be in the range of 299–303 K. A bundle of 230 AO-H fibers, 15 cm in length, was packed in a 1-cm-inner-diameter column of 15 cm length.

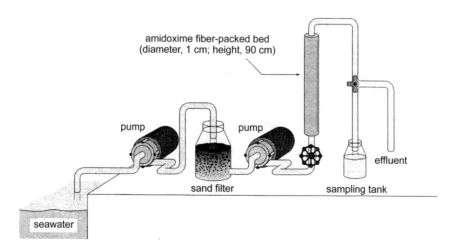

Fig. 4.16 Experimental apparatus placed on the coast of the Pacific Ocean for uranium recovery from seawater. Reprinted with permission from Ref. [62]. Copyright 1991 American Chemical Society

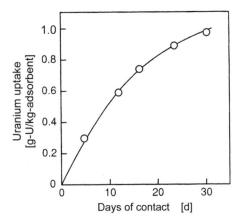

Fig. 4.17 Uranium uptake from seawater versus contact time. Reprinted with permission from Ref. [62]. Copyright 1991 American Chemical Society

Seawater flowed through both the lumen and shell of an AO-H fiber placed in a direction parallel to the flow and did not permeate across the hollow-fiber wall. The void fraction of the bed was calculated to be 76%. Connecting six columns in series afforded a 90-cm-high AO-H bed.

Seawater was allowed to flow over 30 days through the 90-cm-high AO-H bed at a superficial velocity of 4 cm/s, which corresponded to a mean residence time of 22.5 s. The average amount of uranium adsorbed on the AO-H fibers in the bed is shown in Fig. 4.17 as a function of contact time. We achieved a uranium uptake of 0.97 g of U/kg of AO-H fibers and a recovery ratio of 31% after 30 days of contact, where the recovery ratio is defined as the ratio of the total amount adsorbed in the bed to the total amount that passed through the bed. The resulting uranium content of the AO-H fibers is comparable to that of terrestrial low-grade uranium ores.

The distribution of uranium adsorbed in the AO-H fiber-packed bed after 30 days of contact is shown in Fig. 4.18 along with those of magnesium and calcium. Almost flat distributions for the three elements were observed along the bed. From these results, the concentration factor of each element, defined as a ratio of the amount adsorbed on the adsorbent to the concentration in natural seawater, was calculated as 290,000 for uranium, 29 for magnesium, and 49 for calcium.

The adsorption bed can be easily switched to an elution bed by having an eluent flow in place of seawater. First, water was introduced to wash the bed. Second, pre-elution with 0.01 M HCl up to 50 bed-volumes (BV) was necessary to elute almost all of the magnesium and calcium adsorbed on the AO-H fibers. Third, elution with 1 M HCl was carried out. The curve of elution of uranium using 1 M HCl at a superficial velocity 0.0125 cm/s is shown in Fig. 4.19. The uranium concentration in the eluent was 230 g of U/m^3 at the peak at 1.2 BV. The integrated concentration in the eluent collected in the range between 0.7 and 3.1 BV was 45 g of uranium per m^3 for 95% recovery of uranium. Therefore, the concentration ratio in the adsorption and elution systems using the AO-H fiber-packed bed was calculated as 14,000.

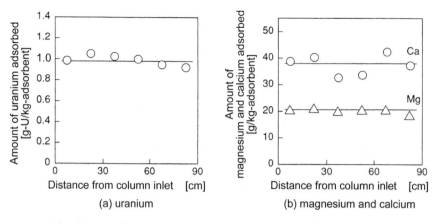

Fig. 4.18 Distribution of the amounts of U, Mg, and Ca adsorbed onto the amidoxime hollow-fiber-packed bed in the direction of its height. Reprinted with permission from Ref. [62]. Copyright 1991 American Chemical Society

Fig. 4.19 Elution curve of uranium from an amidoxime hollow-fiber-packed bed with 1 M HCl. Reprinted with permission from Ref. [62]. Copyright 1991 American Chemical Society

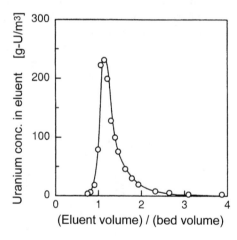

4.1.5 Amidoxime Nonwoven Fabric Submerged in the Pacific Ocean

From a feasibility study, a recovery system using ocean currents and waves were reported to be more advantageous than that using a pump. We employed a nonwoven fabric instead of the AO-H fibers as a packing material for an adsorption cage submerged in the ocean [70]. Acrylonitrile and methacrylic acid were cografted onto a nonwoven fabric made of polyethylene/polypropylene by radiation-induced graft polymerization, and subsequently some cyano groups were converted into amidoxime groups. Cograft polymerization of a hydrophilic vinyl monomer or methacrylic acid with acrylonitrile onto the nonwoven fabric was effective in improving the adsorption rate of uranium onto the resulting amidoxime adsorbent in seawater.

We submerged adsorption cages for uranium recovery in the Pacific Ocean 7 km offshore from Sekine-Hama in Aomori Prefecture, Japan (141° 18.0′E, 41° 24.4′N). The sea depth at the submersion site was approximately 40 m. The adsorption cage, 16 m^2 in cross-sectional area and 16 cm in height, consisted of 144 stacks of the amidoxime adsorbent in the form of nonwoven AO fabric (AO-NWF). Each stack consisted of 120 sheets of the AO-NWF, 29 cm long, 16 cm wide, and 0.2 mm thick, with 59 sheets of spacer nets (Fig. 4.20). The stacks were packed regularly into the adsorption cages in a direction parallel to the perpendicular axis of the adsorption cages. Three adsorption cages connected in series with four ropes were submerged in seawater at a span of 1.5 m by a floating frame that was stabilized by four buoys suspended by four 40-ton anchors placed on the ocean floor. The total mass of dry AO-NWF placed into the three adsorption cages was 350 kg. The frame was designed to be resistant to the following ocean conditions: wind strength, 30 m/s; tidal current, 1.03 m/s; and wave height, 10 m.

Seawater rapidly entered the sheets of the spacer nets upward and downward as induced by the wave motion and slowly penetrated the sheets of AO-NWF. In addition, the tidal motion stirred the seawater horizontally between the adsorption cages. Uranium species in the bulk seawater were transported to the amidoxime groups of the polymer chain grafted onto the nonwoven fabric via both convective and diffusional mass transfer.

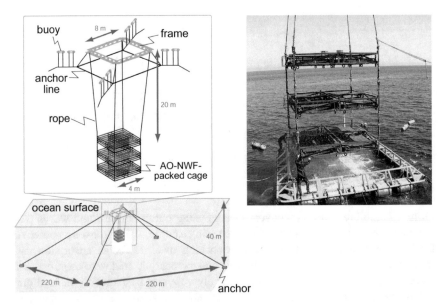

Fig. 4.20 Uranium uptake from seawater using an AO-NWF-packed bed submerged in the Pacific Ocean. Reprinted with permission from Ref. [68]. Copyright 2003 Taylor & Francis Ltd.

The uranium recovery experiment was performed from autumn of 1999 to autumn of 2001. The adsorption cages were pulled out of the seawater using a crane ship every 20–40 days. Throughout this experiment, the total amount of uranium recovered by the 350-kg AO-NWF was >1 kg in terms of yellow cake during a total submersion time of 240 days in the ocean; uranium ores with approximately 0.3% uranium were aquacultured in the Pacific Ocean.

Biofouling was observed on the surface of the stacks. Biofouling occurs owing to the adhesion and subsequent growth of marine microorganisms and algae; however, most of the marine microorganisms were removed by immersing the stacks in fresh water after the stacks were taken out of the adsorption cage. The drastic decrease in ionic strength caused the detachment of these marine microorganisms from the surface of the stacks.

Seawater is regarded as a soup of the ocean floor; therefore, almost all naturally occurring elements dissolve in seawater to form various species and concentrations of ions. For example, a predominant anionic form of chlorine as a major component with a concentration of 0.5 mol/L is chloride ion (Cl^-), whereas uranium, as a minor component with a concentration of 1.4×10^{-8} mol/L, forms complexes with carbonate ions at the pH of seawater in the range of 8.0–8.2 to form the uranyl tricarbonate anion. The molar ratio of chlorine to uranium is 10^9. The amidoxime group selectively captures uranium species in seawater and easily releases the uranium species with 1 M HCl. Functional groups with much higher specificity would capture uranium but not allow a quantitative elution with 1 M HCl. In addition, the much more severe conditions required for the elution of uranium can degrade the functional groups. Functional groups combining selectivity during the adsorption with mild conditions for elution are a requisite for repeated use of the adsorbent.

4.1.6 High-Capacity Capture of Proteins Using Ion-Exchange Fibers Prepared by Radiation-Induced Emulsion Graft Polymerization [71]

On the graft chain, the polymer brush exclusively adsorbs proteins as macro-molecules. In other words, proteins are excluded by the polymer root surrounded by the polymer matrix or trunk polymer. A higher ratio of the polymer brush to the polymer root in the graft chain leads to a higher equilibrium binding capacity of proteins.

To minimize the amount of solvent for the vinyl monomer in radiation-induced graft polymerization, Seko et al. [72, 73] employed the emulsion graft polymerization of glycidyl methacrylate (GMA) onto the irradiated fiber to prepare the adsorbents for metal ions.

To increase the ratio of the polymer brush to the polymer root, we adopted the emulsion graft polymerization of GMA onto polyethylene fiber with a diameter of

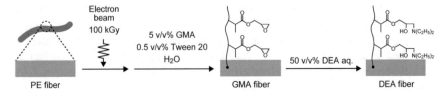

Fig. 4.21 Emulsion graft polymerization of GMA onto polyethylene fiber

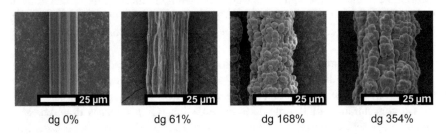

dg 0%   dg 61%   dg 168%   dg 354%

Fig. 4.22 SEM images of GMA-grafted fibers obtained by radiation-induced emulsion graft polymerization

approximately 20 μm. The grafting scheme is shown in Fig. 4.21, where the emulsion of the vinyl monomer consists of GMA, Tween 20, and water at a volume ratio of 5:0.5:94.5. The resultant GMA-grafted fiber is referred to as an emu-GMA (dg) fiber, where dg denotes the degree of GMA grafting. SEM images of the emu-GMA(dg) fibers with values of dg are shown in Fig. 4.22. With increasing degree of GMA grafting, the surface of the polyethylene fiber appeared to be coated with an increasing amount of polymer in a globular form, which corresponds to the polymer brush.

A diethylamino (DEA) group was successfully introduced into the emu-GMA(dg) fiber at a molar conversion of the epoxy group into the DEA group of 93% up to a dg value of 350% to capture proteins via an anion-exchange interaction. Bovine serum albumin solution flowed through the bed charged with the emu-DEA(61) fiber derived from the emu-GMA(61) fiber, and the breakthrough curves of a fiber-packed bed with an inner diameter of 0.5 cm and a height of 2 cm for BSA were determined at various SVs. A similar experiment was conducted with a bed charged with DEAE-Sepharose-Fast Flow beads with an average diameter of 90 μm (GE Healthcare Co.). The DBCs of the emu-DEA(61)-fiber- and DEAE-Sepharose-FF-bead-packed beds are shown in Fig. 4.23. The DBC of the emu-DEA(61)-fiber-packed bed was clearly higher than that of the DEAE-Sepharose-bead-packed bed up to an SV of 1200 h^{-1}. This is because the diffusion path of BSA to the DEA group of the polymer brush immobilized onto the polyethylene fiber is shorter than that of the BSA into the beads. The application of the emulsion graft polymerization of AAc to polyethylene fiber was effective in improving the protein binding capacity of the weakly basic anion-exchange fiber.

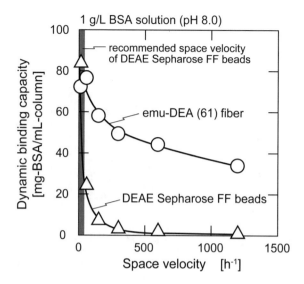

Fig. 4.23 Comparison of dynamic binding capacity between emu-DEA(61)-fiber- and DEAE-Sepharose-FF-bead-packed columns. Reprinted with permission from Ref. [82]. Copyright 2013 Japan Society of Ion Exchange

4.2 Nonporous Films

Dialysis is an application of the principle that a solute is transported through an artificial or natural membrane driven by the transmembrane differences of concentration and electropotential. Dialysis membranes must have interstices larger than the molecules or ions that diffuse through them. When ions permeate the membranes, they can interact with charges immobilized on the membranes. Thus, the size and charge of the permeate govern the flux, which is defined as the permeation rate divided by the membrane area.

Dialysis induced by a concentration difference across the membrane is referred to as diffusion dialysis. In contrast, in electrodialysis, ion transport is driven by the electropotential difference across the membrane. In salt manufacturing, salt concentration is increased by a factor of approximately seven by electrodialysis using cation- and anion-exchange membranes. The apparatus for laboratory-scale electrodialysis is illustrated in Fig. 4.24. During the electrodialysis, ions are transported through interstices between the charged polymer chains.

In Japan, edible salt (sodium chloride) has been manufactured by the electrodialysis of seawater since 1979. In 2013, the annual production and consumption rates of sodium chloride were 9.3×10^5 and 8.0×10^6 tons, respectively. Currently used ion-exchange membranes are produced by a paste method: A base cloth is immersed in a mixture of vinyl monomers, followed by polymerization and introduction of ion-exchange groups. Furthermore, a coating is used to enable the

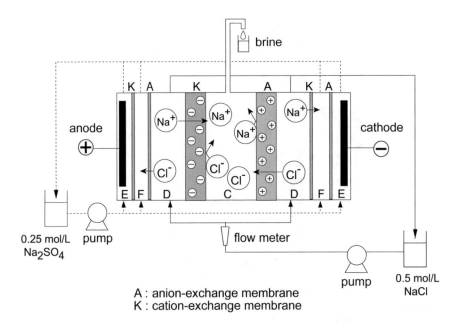

	width [mm]	cross-sectional area [cm²]
concentration chamber	1.5	8.0 (2.0 × 4.0 cm)
desalination chamber	6.5	8.0 (2.0 × 4.0 cm)
electrode chamber	6.5	8.0 (2.0 × 4.0 cm)
wash chamber	6.5	8.0 (2.0 × 4.0 cm)

Fig. 4.24 Salt enrichment of seawater using an electrodialyzer

selective permeation of monovalent ions over divalent ions such as magnesium, calcium, and sulfate ions. We developed novel ion-exchange membranes for the electrodialysis of seawater by radiation-induced graft polymerization in collaboration with the Research Institute of Salt and Sea Water Science of The Salt Industry Center of Japan.

4.2.1 Formation of Interstices in Films [74–81]

Novel ion-exchange membranes that allow the selective permeation of monovalent ions may be prepared by radiation-induced graft polymerization from commercially available polymeric films. First, radicals are produced uniformly over a polymeric film by irradiation with an electron beam. Second, the irradiated film is immersed in vinyl monomer solutions to initiate graft polymerization. As shown in Fig. 4.25a, when an epoxy-group-containing vinyl monomer is used, epoxy group may be converted into ion-exchange groups such as sulfonic acid groups ($-SO_3H$) and trimethylammonium groups ($-N(CH_3)_3$). As shown in Fig. 4.25b, films grafted with ion-exchange-group-containing vinyl monomers are ion-exchange membranes.

Low electrical resistance of ion-exchange films is a requirement for their application to salt manufacturing by electrodialysis. To achieve this, ion-conducting

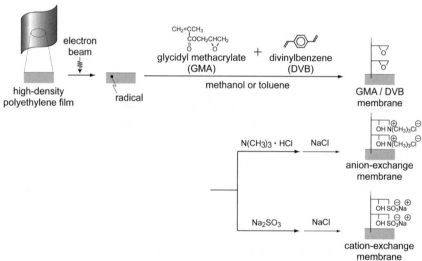

(a) grafting of epoxy-group-containing monomers and subsequent conversion of the epoxy groups into ion-exchange groups

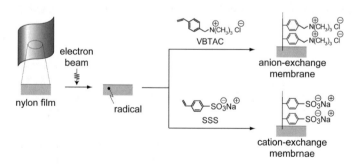

(b) grafting monomers containing ion-exchange groups

Fig. 4.25 Introduction of anion- and cation-exchange groups into commercially available polymeric films

Table 4.1 Commercially available polymeric films for preparation of ion-exchange membranes by radiation-induced graft polymerization

Trunk polymer		Film thickness (µm)	Film density (g/cm^3)
Polyethylene	LDPE	30	0.812
	HDPE	35	0.940
	UMWPE	30	0.933
Polyacrylonitrile	PAN	30	1.14
Polyamide	nylon 6	15	1.19
Aromatic compounds	PET	30	1.22
	PEN	25	1.36
Fluorine compounds	ETFE	25	1.74
	PFA	50	2.21

LDPE low-density polyethylene; *HDPE* high-density polyethylene; *UHMWPE* ultrahigh molecular weight polyethylene; *PAN* polyacrylonitrile; *PET* polyethylene terephthalate; *PEN* polyethylene naphthalate; *ETFE* ethylene-tetrafluoroethylene copolymer; *PFA* tetrafluoroethylene-perfluoroalkyl vinyl ether copolymer

channels that penetrate the film must be formed so that sodium and chloride ions can permeate the membrane. An electron beam of sufficient energy is used to produce radicals uniformly throughout the initial film. Furthermore, the solvent for vinyl monomers is selected to graft the chains uniformly over the film thickness. The temperature and concentration of the vinyl monomers were optimized in terms of electrical resistance and ion-exchange capacity.

Nine kinds of commercially available polymer films were adopted as listed in Table 4.1 along with their thicknesses and densities. The ion-exchange membranes were prepared in accordance with the schemes shown in Fig. 4.18. The combination of a trunk polymer and a vinyl monomer that satisfied the required mechanical strength and ion-exchange capacity was selected. In addition, the properties necessary for salt manufacturing, i.e., chloride ion concentration in the chamber and electrical resistance, were determined in a laboratory-scale dialysis cell using 0.5 M sodium chloride as a seawater model. As a result, the following two versions of ion-exchange membranes comparable to currently used ion-exchange membranes were selected:

(1) GMA dissolved in toluene was graft-polymerized onto a previously irradiated high-density polyethylene film (HDPE) with a thickness of 35 µm, followed by the conversion of epoxy groups into sulfonic acid (SS) and trimethylammonium (TMA) groups. The resultant membrane is referred to as the TMA-HDPE membrane.

(2) SSS and VBTAC dissolved in water were graft-polymerized onto a previously irradiated nylon-6 film (NY) with a thickness of 25 µm. The resultant membrane is referred to as the VBTAC-NY membrane.

4.2.2   Characterization of Graft-Type Ion-Exchange Membranes Used in Electrodialysis

By radiation-induced graft polymerization, new ion-exchange membranes were derived from commercially available HDPE and NY films. Their performance in the electrodialysis of seawater was evaluated as a function of the degree of grafting (dg), where dg is defined as the percentage of mass gained via graft polymerization (Chaps. 1 and 2). First, ion-exchange groups must be distributed uniformly throughout the membrane because a nonuniform distribution will induce problems with electrical and mass-transfer characteristics. If the values of dg were over 70% for HDPE films or VBTAC or NY films, then the functionality was introduced uniformly across the membrane, as shown in Fig. 4.26. A map representing the relationship between chloride ion concentration in a concentration chamber and the electrical resistance of the membrane is shown in Fig. 4.27. Current lines have been inserted in these plots. The domain to the left and above the current line is favorable. The anion-exchange membrane that was prepared by cografting GMA and DVB followed by the introduction of trimethylammonium groups exhibited higher performance than the current membrane shown by solid points in this figure: The crosslinked graft membrane was SELEMION™ ASA, which was manufactured by AGC Engineering Co. Researchers at the Seawater Research Institute succeeded in preparing a membrane with much higher performance than the currently used membrane.

On a laboratory scale, preparation by pre-irradiation grafting is advantageous over the current pasting method in that radicals are formed using a dry process and the reaction is simpler. The membrane size of 2×4 cm^2 in the laboratory device must be scaled-up to 110×230 cm^2 on a plant scale. The problems accompanying the scale-up are addressed in relation to the diaphragm and nonwoven fabric manufactured by radiation-induced graft polymerization (Chap. 6).

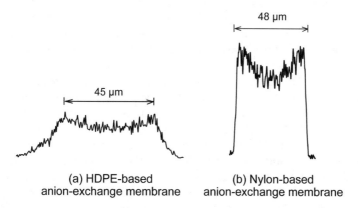

(a) HDPE-based
anion-exchange membrane

(b) Nylon-based
anion-exchange membrane

Fig. 4.26 Distributions of chloride ions bound to trimethylammonium groups across anion-exchange membranes

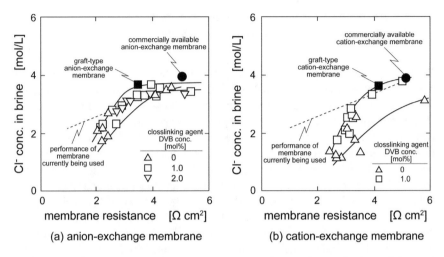

Fig. 4.27 Chloride ion concentration in brine versus membrane resistance in electrodialyzer

4.2.3 Structural Changes of Polyethylene Films [80]

Characterization of the graft-type ion-exchange membranes is useful for improving their performance. Structural changes of the HDPE film accompanied by graft polymerization of GMA and introduction of ion-exchange groups were determined by the following four techniques: X-ray diffraction (XRD), differential scanning calorimetry (DSC), small-angle X-ray scattering (SAXS), and small-angle neutron scattering (SANS). The latter two techniques are favorable in that the measurements are applicable to wet films.

The size of crystallites, the distance between the chains of folded polyethylene, the degree of crystallinity, and the size of the regular structure can be determined. Although these parameters cannot provide quantitative information on ion-conduction channels through which ions penetrate the grafted polymer chain such as their width, length, and shape, the ion-conducting domain can be identified by determining the formation site of the graft chain and the profile of the functional groups.

First, the HDEHP film as a starting polymeric material includes crystalline and amorphous domains (Fig. 4.28). The crystalline domain is made of lamella consisting of folded polyethylene chains. Irradiation produces radicals uniformly over the HDPE film. The GMA-grafted HDPE film with a dg of 100% was characterized as follows: The distance between folded PE chains remained constant after GMA grafting, whereas the distance between the crystallites and the size of the crystallites decreased. When the epoxy groups of the graft chain were converted into trimethylammonium groups at a molar conversion of 60%, the distance between the folded PE chains and the crystal size did not change while the distance between the crystals increased.

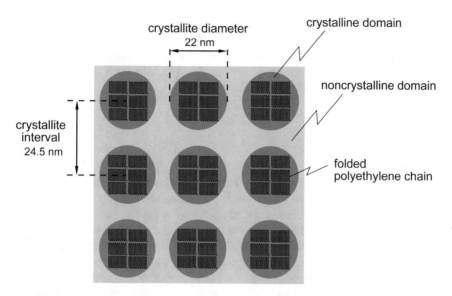

crystallite diameter
22 nm

crystalline domain

noncrystalline domain

crystallite
interval
24.5 nm

folded
polyethylene chain

Fig. 4.28 Illustrations of the inner structure of a high-density polyethylene film

The radicals were uniformly produced, but the vinyl monomers could not invade the inside of the crystalline domains because their size was larger than the distance between the folded PE chains. Although the radicals inside the crystalline part are not directly involved in the graft polymerization, they diffuse to the crystallite surfaces while drawing hydrogen along the PE chain to contribute to the graft polymerization. The radicals on the crystals and in the amorphous domains initiate the graft polymerization, and the polymer chain grows.

As the graft polymerization proceeds, the distance between the crystallites increases and the crystal size decreases owing to the collapse around the periphery of the crystal. Subsequently, the introduction of ion-exchange groups causes the intake of hydrated water, which increases the distance between the crystallites. However, the crystal size remains constant because the graft chain is not located in the crystallite. These structural changes of PE are reasonable from the viewpoint of the invasion of chemicals used in the modifications.

Studies of the modification of various commercially available polymeric films into ion-exchange membranes for electrodialysis by radiation-induced graft polymerization have been detailed in this chapter. Originally, the initial polymeric films possess interstices through which gaseous molecules can permeate. To prepare an ion-exchange membrane for electrodialysis, the interstices must be capable of being permeated by hydrated ions, i.e., sodium and chloride ions, and the membrane must have a low electrical resistivity against reverse diffusion driven by the concentration differences across the membrane. Electron-beam-induced graft polymerization is advantageous compared with the conventional pasting method in that an electron beam can readily produce radicals uniformly throughout the film. The project on the replacement of conventional ion-exchange membranes with novel ion-exchange membranes is promising [82], as is described in Chap. 6. The preparation scheme discussed here is useful for other applications [83, 84] such as fabrication of ion-exchange membranes for fuel cells.

4.3   Particles [85–88]

One of macroscopic advantages of radiation-induced graft polymerization is that it is applicable to arbitrary forms of trunk polymers: Most commercially available polymers designed for separation and reaction are in the form of beads of various sizes. Consequently, we have intensively studied the development of functional polymers of various forms other than the bead form by radiation-induced graft polymerization and subsequent chemical modifications. We succeeded in preparing ion-exchange and extractant-impregnated particles with the highest resolution ever when the particles were packed in a column for elution chromatography of proteins and rare-earth metal ions, respectively.

4.3.1   Elution Chromatography of Proteins

Preparation schemes for ion-exchange particles by radiation-induced graft polymerization are shown in Fig. 4.29. First, polyethylene particles were irradiated with an electron beam. Second, GMA as a precursor monomer was graft-polymerized onto the irradiated polyethylene particles. Third, epoxy groups were converted into sulfonic acid and diethylamino groups by reaction with sodium sulfite (SS) and diethylamine (DEA), respectively. The resultant ion-exchange particles are referred to as SS and DEA particles, respectively.

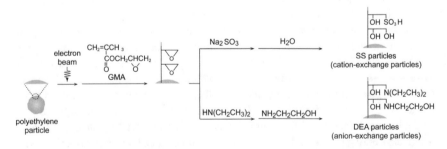

Fig. 4.29 Preparation schemes for ion-exchange particles via radiation-induced graft polymerization and subsequent chemical modifications

Ion-exchange beads have been widely used in food, pharmaceutical, and water-treatment industries. In particular, GE Healthcare and Merck Co. have manufactured ion-exchange beads for protein purification under the brand names, SOURCE and Tentacle, respectively. In contrast, our ion-exchange particles are not spherical and have a relatively wide size distribution. The ion-exchange groups are concentrated at the periphery of the particles. Thus, we anticipated that our particles would have a disadvantage compared with commercially available ion-exchange beads.

Prescribed amounts of three proteins were loaded on the top end of an ion-exchange particle-packed column before gradient elution. Sodium chloride solution, the concentration of which was increased from 0 to 1.5 M, flowed down the column, and the amount of the proteins in the effluent were determined continuously. The resolution of the elution chromatogram for graft-type particles was compared with that of the commercially available ion-exchange bead SOURCE 30S.

SS particles derived from 60-μm-diameter polyethylene particles were packed into a column with an inner diameter of 5 mm and a packed height of 1.5 cm. On the other hand, SOURCE 30S was packed into a similar column to a packed height of 2.5 cm to produce the same pressure loss across the column as the SS particles.

A mixture of three proteins, i.e., α-chymotrypsinogen (pI 9.2, Mr 25,000), cytochrome C (pI 10.5, Mr 12,400), and lysozyme (pI 11.2, Mr 14,700), was employed as a typical protein solution at a mass ratio of 3:4:3. An identical amount of the mixture, 0.4 mg, was loaded onto the SS-particle- and SOURCE 30S-packed columns, followed by gradient elution with sodium chloride solution at linear velocities (LVs) ranging from 150 to 600 cm/h. As shown in Fig. 4.30, the SS column did not result in an overlap of peak tails of the proteins. Furthermore, higher resolution at higher loading at the same LV was achieved.

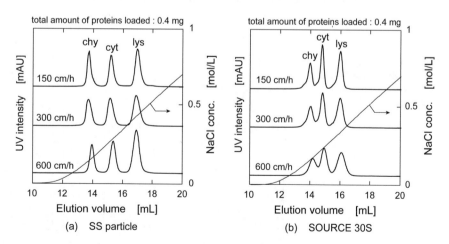

Fig. 4.30 Comparison of elution chromatograms of proteins at various linear velocities between (**a**) SS-particle- and (**b**) SOURCE 30S-packed beds

A column charged with DEA particles also exhibited excellent performance in elution chromatography of proteins, as described in Chap. 5, because the proteins were entangled by the graft chains via electrostatic interactions to form multilayer binding structures. Enhanced ion-exchange binding enables the proteins to magnify the differences in their binding strengths, resulting in high resolution with the sodium chloride solutions. This effect compensates for any disadvantage of particles over beads in terms of size distribution.

4.3.2 Elution Chromatography of Rare-Earth Metal Ions

Rare-earth-metal (REM)-containing ores are usually eluted with sulfuric acid, and REM ions are extracted by liquid–liquid extraction prior to concentration and purification. Industrial equipment designed for liquid–liquid extraction includes mixer-settlers and rotating disk columns (RDCs) (Fig. 4.31). Extractants with high affinity for REMs are dissolved into organic solvents and put in contact with aqueous phases containing REM ions. REM ions are then extracted with extractants at the interface between the organic and aqueous phases. This procedure is referred to as forward extraction. Subsequently, REM ions extracted into the organic phase are re-extracted with an aqueous phase. This procedure is backward extraction. Through this method of forward and backward extractions, competing species are rejected and target species are enriched.

From an environmental viewpoint, the replacement of organic solvents with hydrophobic polymeric materials has been proposed for solid-phase extraction. Instead of dissolving extractants into organic solvents, we impregnated the extractants onto the hydrophobic moieties of graft chains such as a dodecyl group ($-C_{12}H_{25}$), prepared by radiation-induced graft polymerization and subsequent

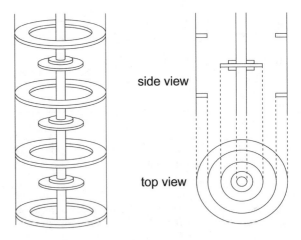

Fig. 4.31 Rotating disk column (RDC): plant-scale equipment for liquid–liquid extraction

side view

top view

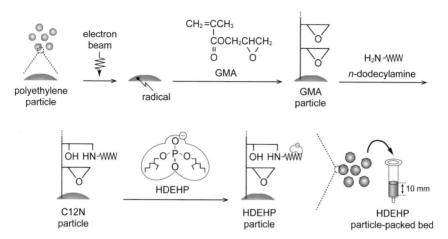

Fig. 4.32 Impregnation of acidic extractant HDEHP into polymer chains grafted onto PE particles

chemical modifications. The hydrophobic graft chains behave like the organic phase for the extractants.

The most powerful permanent magnet, a neodymium magnet, consists of iron (Fe), neodymium (Nd), dysprosium (Dy), and boron (B), which is an invention of a Japanese engineer, Dr. Masato Sagawa. We initiated the recovery of Nd and Dy from powder from the cutting of the neodymium magnet. Similarly to the mixture of three proteins separated with a cation-exchange particle-packed column by elution chromatography, the mixture of neodymium and dysprosium ions was separated with an HDEHP-impregnated particle-packed column by elution chromatography.

An impregnation scheme for HDEHP onto polyethylene particles is shown in Fig. 4.32. Particles with an impregnation density of 0.25 mol/kg were packed into a column with an inner diameter of 8.8 cm and a packing height of 10 cm. A mixture of neodymium and dysprosium ions was loaded on the top end of the column at a mass ratio of Nd to Dy of 29–4. Subsequently, two solutions of hydrochloric acid, i.e., concentrations of 0.2 and 1.5 mol/L, were forced to flow downward in sequence through the column. The two rare-earth metal ions in the effluent flowing out at the bottom of the column were identified. As shown in Fig. 4.33, the two rare-earth metal ions were completely separated.

Hydrophobic ligands may be introduced into the polymer chains grafted onto the periphery of the polyethylene particles and then extractants may be impregnated onto the hydrophobic ligands. The hydrophobic graft phase behaves like an organic phase. Extractants dissolved in the graft phase provide high resolution in elution chromatography. In addition, a short diffusional mass-transfer path and high mobility of extractants in the graft phase provide high performance in separation. A comparison between the graft-type extractant-impregnated particles and commercially available extractant-impregnated beads is detailed in Chap. 5.

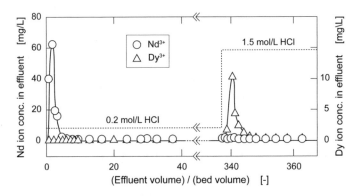

Fig. 4.33 Elution chromatogram of HDEHP-impregnated particle-packed bed for Nd^{3+} and Dy^{3+} with 0.2 and 1.5 M HCls

The use of porous hollow-fiber membranes and porous sheets as trunk polymers for grafting was described in Chap. 3. This chapter dealt with fibers, films, and particles as trunk polymers for grafting. In the next chapter, the competition between graft-type adsorbents and conventional adsorbents is described. Examples of the products prepared by radiation-induced graft polymerization are detailed in Chap. 6.

References

1. K. Saito, Preparation of polymeric fibers immobilizing inorganic compounds, enzymes, and extractants designed for radionuclide decontamination, ultrapure water production, and rare-earth metal purification. Kobunshi Ronbunshu **71**, 211–222 (2014)
2. G.B. Barton, J.L. Hepworth, E.D. McClanahan, R.L. Moore, H.H. Van Tuyl, Chemical processing waster; recovering fission products. Ind. Eng. Chem. **50**, 212–216 (1958)
3. J. Letho, R. Harjula, J. Wallace, Adsorption of cesium on potassium cobalt hexacyanoferrate (lll). J. Radioanal. Nucl. Chem. **111**, 297–304 (1987)
4. H. Mimura, J. Lehto, R. Harjula, Chemical and thermal stability of potassium nickel hexacyanoferrate(ll). J. Nucl. Sci. Technol. **34**, 582–587 (1997)
5. I. Izmail, M. El-Sourough, N. Moneim, H. Aly, Equilibrium and kinetics studies of the sorption of cesium by potassium nickel hexacyanoferrate complex. J. Radioanal. Nucl. Chem. **240**, 59–67 (1999)
6. K. Watari, M. Izawa, Separation of radiocesium by copper ferrocyanide-anion exchange resin. J. Nucl. Sci. Technol. **2**, 321–322 (1965)
7. K. Watari, K. Imai, M. Izawa, Radiochemical application of iron ferrocyanide-anion exchange resin. J. Nucl. Sci. Technol. **5**, 309–312 (1968)
8. T.P. Valsala, A. Joseph, J.G. Shah, K. Raj, V. Venugopal, Synthesis and characterization of cobalt ferrocyanides loaded on organic anion exchanger. J. Nucl. Mater. **384**, 146–152 (2009)
9. T.P. Valsala, S.C. Roy, J.G. Shah, J. Gabriel, K. Raj, V. Venugopal, Removal of radioactive caesium from low level radioactive waste (LLW) streams using cobalt ferrocyanide impregnated organic anion exchanger. J. Hazard Mater. **166**, 1148–1153 (2009)

10. K. Tanihara, Selective separation of cesium from strongly acidic nitrate media by repeated use of cupric ferrocyanide-silica gel composite ion exchanger of redox type. in *Reports of the Kyushu National Industrial Research Institute*, No. 61, pp. 23–28 (1998)

11. H. Mimura, M. Kimura, K. Akiba, Y. Onodera, Selective removal of cesium from highly concentrated sodium nitrate neutral solutions by potassium nickel hexacyanoferrate(ll)-loaded silica gels. Solvent Extr. Ion Exch. **17**, 403–417 (1999)

12. H. Mimura, M. Kimura, K. Akiba, Y. Onodera, Separation of cesium and strontium by potassium hexacyanoferrate(ll)-loaded zeolite A. J. Nucl. Sci. Technol. **36**, 307–310 (1999)

13. R. Ishihara, K. Fujiwara, T. Harayama, Y. Okamura, S. Uchiyama, M. Sugiyama, T. Someya, W. Amakai, S. Umino, T. Ono, A. Nide, Y. Hirayama, T. Baba, T. Kojima, D. Umeno, K. Saito, S. Asai, T. Sugo, Removal of cesium using cobalt-ferrocyanide-impregnated polymer-chain-grafted fibers. J. Nucl. Sci. Technol. **48**, 1281–1284 (2011)

14. Y. Hirayama, Y. Okamura, K. Fujiwara, T. Sugo, D. Umeno, K. Saito, Effect of salt concentration of cesium solution on cesium-binding capacity of potassium cobalt-hexacyanoferrate-impregnated fiber. *J. Chem. Eng. Japan*. **39**, 28–32 (2013)

15. Y. Okamura, K. Fujiwara, N. Iijima, T. Syoda, K. Suzuki, T. Sugo, T. Shimidu, R. Itagaki, A. Takahashi, T. Ono, T. Kikuchi, T. Someya, R. Ishihara, T. Kojima, D. Umeno, K. Saito, J. Ion Exch. **24**, 8–13 (2013)

16. W. Amakai, M. Sugiyama, K. Fujiwara, T. Sugo, D. Umeno, K. Saito, Adsorption isotherms for cesium ions in seawater of insoluble cobalt and nickel ferrocyanide-impregnated fibers. Bull. Soc. Sea Water Sci. Jpn. **68**, 18–24 (2014)

17. W. Amakai, Y. Okamura, K. Fujiwara, T. Sugo, D. Umeno, K. Saito, Impregnation of insoluble cobalt ferrocyanide onto poly-(vinylbenzyl) trimethylammonium-chloride chain grafted onto 6-nylon fiber for the removal of cesium ions from freshwater. J. Soc. Remedi. Radioact. Contami. Environ. **2**, 93–99 (2014)

18. S. Goto, W. Amakai, K. Fujiwara, T. Sugo, T. Kojima, S. Kawai-Noma, D. Umeno, K. Saito. A novel preparation scheme for cesium-adsorptive fibers to raise the impregnation percentage of insoluble cobalt ferrocyanide. Bull. Soc. Sea Water Sci. Jpn. **68**, 298–304 (2014)

19. T. Someya, S. Asai, K. Fujiwara, T. Sugo, D. Umeno, K. Saito. Removal of cesium ions from contaminated seawater in closed area using adsorptive fiber. Bull. Soc. Sea Water Sci. Jpn. **69**, 42–48 (2015)

20. M. Sugiyama, S. Goto, T. Kojima, K. Fujiwara, T. Sugo, D. Umeno, K. Saito, Impregnation process of insoluble cobalt ferrocyanide onto anion-exchange fiber prepared by radiation-induced graft polymerization. Radioisotopes **64**, 219–228 (2015)

21. S. Goto, S. Umino, W. Amakai, K. Fujiwara, T. Sugo, T. Kojima, S. Kawai-Noma, D. Umeno, K. Saito, Impregnation structure of cobalt ferrocyanide microparticles by the polymer chain grafted onto nylon fiber. J. Nucl. Sci. Technol. **53**, 1251–1255 (2016)

22. J. Letho, A. Clearfield, The ion exchange of strontium on titanate $Na_4Ti_9O_{20} \cdot XH_2O$. J. Radioanal. Nucl. Chem. **118**, 1–13 (1987)

23. T. Harayama, S. Umino, S. Uchiyama, M. Sugiyama, K. Fujiwara, T. Sugo, S. Asai, T. Kojima, D. Umeno, K. Saito, Preparation of adorptive fibers for removal of strontium from seawater. Bull. Soc. Sea Water Sci. Jpn. **66**, 295–300 (2012)

24. S. Umino, M. Kono, K. Fujiwara, T. Sugo, S. Kawai-Noma, D. Umeno, K. Saito, Selection of scheme for impregnation of sodium titanate onto ion-exchange fibers for radioactive strontium removal from seawater. Bull. Soc. Sea Water Sci. Jpn. **68**, 89–93 (2014)

25. Y. Nakatani, S. Umino, M. Sugiyama, K. Fujiwara, T. Sugo, T. Kojima, D. Umeno, K. Saito, Impregnation of hydrous titanium oxide onto cation-exchange polymer chain grafted onto nylon fiber. Bull. Soc. Sea Water Sci. Jpn. **68**, 196–201 (2014)

26. M. Kono, S. Umino, K. Fujiwara, T. Sugo, T. Kojima, D. Umeno, K. Saito, Repeated deposition of titanium compounds onto 6-nylon fiber for removal of strontium from seawater. Bull. Soc. Sea Water Sci. Jpn. **68**, 258–263 (2014)

27. M. Kono, S. Umino, S.I. Goto, K. Fujiwara, T. Sugo, T. Kojima, S. Kawai-Noma, D. Umeno, K. Saito, Preparation of adsorptive fiber by a combination of peroxo complex of titanium

anion and DMAPAA-grafted fiber for the removal of strontium from seawater. Bull. Soc. Sea Water Sci. Jpn. **69**, 90–97 (2015)

28. M. Katagiri, M. Kono, S.I. Goto, K. Fujiwara, T. Sugo, S. Kawai-Noma, D. Umeno, K. Saito, Impregnation of sodium titanate onto DMAPAA-grafted fiber under mild reaction conditions and its strontium removal performance from seawater. Bull. Soc. Sea Water Sci. Jpn. **69**, 270–276 (2015)

29. S. Naruke, S.I. Goto, M. Katagiri, K. Fujiwara, T. Suto, S. Kawai-Noma, D. Umeno, K. Saito, Determination of composition and strontium-binding ratio of sodium titanate impregnated onto DMAPAA-grafted fiber. Bull. Soc. Sea Water Sci. Jpn. **70**, 364–368 (2016)

30. T. Yoshikawa, D. Umeno, K. Saito, T. Sugo, High-performance collection of palladium ions in acidic media using nucleic-acid-base-immobilized porous hollow-fiber membranes. J. Membr. Sci. **307**, 82–87 (2008)

31. T. Sasaki, K. Fujiwara, T. Sugo, S. Kawai-Noma, D. Umeno, K. Saito, Ruthenium removal from water using nucleic-acid base-immobilized fibers. Bull. Soc. Sea Water Sci. Jpn. **69**, 98–104 (2015)

32. M. Sugiyama, K. Ikeda, D. Umeno, K. Saito, T. Kikuchi, K. Ando, Removal of urea from water using urease-immobilized fibers. J. Chem. Eng. Jpn. **46**, 509–513 (2013)

33. S. Kawashima, M. Sugiyama, K. Fujiwara, T. Sugo, T. Kikuchi, F. Koide, H. Kanoh, S. Kawai-Noma, D. Umeno, K. Saito, Preparation of catalase-immobilized and palladium-impregnated fibers for rapid decomposition of hydroperoxide in water. Radioisotopes **64**, 501–507 (2015)

34. K. Saito, T. Miyauchi, Chemical forms of uranium in artificial seawater. J. Nucl. Sci. Technol. **19**, 145–150 (1982)

35. K. Saito, T. Miyauchi, Diffusivity of uranium in artificial seawater. Kagaku Kogaku Ronbunsyu **7**, 545–548 (1981)

36. R.V. Davies, J. Kennedy, T.W. Mciloy, R. Spence, K.M. Hill, Extraction of uranium from seawater. Nature **203**, 1110–1115 (1964)

37. N. Ogata, Review on recovery of uranium from seawater. Bull. Soc. Sea Water Sci. Jpn. **34**, 3–12 (1980)

38. M. Kanno. MMAJ project for the extraction of uranium from seawater. in *Proceedings of and International Meeting Recovery Uranium from Seawater*, vol. 12 (1983)

39. H.J. Schenk, L. Astheimer, E.G. Witte, K. Schwochau, Development of sorbers for the recovery of uranium from seawater. Part 1. Assessment of key parameters and screening studies of sorber materials. Sep. Sci. Technol. **17**, 1293–1308 (1982)

40. L. Astheimer, H.J. Schenk, E.G. Witte, K. Schwochau, Development of sorbers for the recovery of uranium from seawater. Part 2. The accumulation of uranium from seawater by resins containing amidoxime and imidoxime functional groups. Sep. Sci. Technol. **18**, 307–339 (1983)

41. H. Nobukawa, M. Tamehiro, M. Kobayashi, H. Nakagawa, J. Sakakibara, N. Takagi, Development of floating type-extraction system of uranium from sea water using sea water current and wave power. 1. J. Shipbuild. Soc. Jpn. **165**, 281–292 (1989)

42. H. Nobukawa, J. Michimoto, M. Kobayashi, H. Nakagawa, J. Sakakibara, N. Takagi, M. Tamehiro, Development of floating type-extraction system of uranium from sea water using sea water current and wave power. 2. J. Shipbuild. Soc. Jpn., **168**, 319–328 (1990)

43. H. Nobukawa, M. Kitamura, M. Kobayashi, H. Nakagawa, N. Takagi, M. Tamehiro. Development of floating type-extraction system of uranium from sea water using sea current and wave power. 3. J. Shipbuild. Soc. Jpn. **172**, 519–528 (1992)

44. H. Egawa, H. Harada. Recovery of uranium from sea water by using chelating resins containing amidoxime groups. Nippon Kagaku Kaishi. 958–959 (1979)

45. H. Egawa, H. Harada, T. Nonaka. Preparation of adsorption resins for uranium in seawater. *Nippon Kagaku Kaishi*, **11** 1767–1772 (1980)

46. H. Egawa, H. Harada, T. Shuto. Recovery of uranium from sea water by the use of chelating resins containing amidoxime groups. Nippon Kagaku Kaishi. 1773–1776 (1980)

47. H. Egawa, N. Kabay, S. Saigo, T. Nonaka, T. Shuto, Low-cross-linked porous chelating resins containing amidoxime groups. Bull. Soc. Sea Water Sci. Jpn. **45**, 324–332 (1991)
48. T. Hirotsu, S. Katoh, K. Sugasaka, M. Seno, T. Itagaki, Adsorption equilibrium of uranium from aqueous $[UO_2(CO_3)_3]^{4-}$ solutions on a polymer bearing amidoxime groups. J. Chem. Soc. Dalton Trans. **9**, 1983–1986 (1986)
49. S. Katoh, K. Sugasaka, K. Sakane, N. Takai, H. Takahashi, Y. Umezawa, K. Itagaki, Preparation of fibrous adsorbent containing amidoxime group and adsorption property for uranium. Nippon Kagaku Kaishi. 1449–1454 (1982)
50. S. Katoh, K. Sugawaka, K. Sakane, N. Takai, H. Takahashi, Y. Umezawa, K. Itagaki, Enhancement of the adsorptive property of amidoxime-group-containing fiber by alkaline treatment. Nippon Kagaku Kaishi. 1455–1459 (1982)
51. Y. Kobuke, I. Tabushi, T. Aoki, T. Kamaishi, I. Hagiwara, Composite fiber adsorbent for rapid uptake of uranyl from seawater. Ind. Eng. Chem. Res. **27**, 1461–1466 (1988)
52. Y. Kobuke, H. Tanaka, H. Ogoshi, Imidedioxime as a significant component in so-called amidoxime resin for uranyl adsorption from seawater. Polym. J. **22**, 179–182 (1990)
53. T. Saito, S. Brown, S. Chatterjee, J. Kim, C. Tsouris, R.T. Mayes, L.-J. Kuo, G. Gill, Y. Oyola, C.J. Janke, S. Dai, Uranium recovery from seawater: development of fiber adsorbents prepared via atom-transfer radical polymerization. J. Mater. Chem. A **2**, 14674–14681 (2014)
54. S. Brown, Y. Yue, L.-J. Kuo, N. Mehio, M. Li, G. Gill, C. Tsouris, R.T. Mayes, T. Saito, S. Dai, Uranium adsorbent fibers prepared by atom-transfer radical polymerization (ATRP) from poly(vinyl chloride)-co-chlorinated poly(vinyl chloride) (PVC-co-CPVC) fiber. Ind. Eng. Chem. Res. **55**, 4139–4148 (2016)
55. T. Hori, K. Saito, S. Furusaki, T. Sugo, J. Okamoto, Synthesis of a hollow fiber type porous chelating resin containing the amidoxime group by radiation-induced graft polymerization for the uranium recovery. Nippon Kagaku Kaishi. 1792–1798 (1986)
56. K. Saito, S. Yamada, S. Furusaki, T. Sugo, J. Okamoto, Characteristics of uranium adsorption by amidoxime membrane synthesized by radiation-induced graft polymerization. J. Membr. Sci. **34**, 307–315 (1987)
57. T. Hori, K. Saito, S. Furusaki, T. Sugo, J. Okamoto, Adsorption equilibrium of uranium from seawater on chelating resin containing amidoxime group. Kagaku Kogaku Ronbunsyu **13**, 795–800 (1987)
58. K. Saito, T. Hori, S. Furusaki, T. Sugo, J. Okamoto, Porous amidoxime-group-containing membrane for the recovery of uranium from seawater. Ind. Eng. Chem. Res. **26**, 1977–1981 (1987)
59. T. Hori, K. Saito, S. Furusaki, T. Sugo, J. Okamoto. The effect of alkaline and acidic treatment of the properties of amidoxime resin synthesized by radiation-induced graft polymerization. Nippon Kagaku Kaishi. 1607–1611 (1988)
60. K. Uezu, K. Saito, T. Hori, S. Furusaki, T. Sugo, J. Okamoto, Performance of fixed-bed charged with chelating resin of capillary fiber form for recovery of uranium from seawater. Nihon Genshiryoku Gakkaishi **30**, 359–364 (1988)
61. K. Saito, K. Uezu, T. Hori, S. Furusaki, T. Sugo, J. Okamoto, Recovery of uranium from seawater using amidoxime hollow fibers. AIChE J. **34**, 411–416 (1988)
62. K. Uezu, K. Saito, S. Furusaki, T. Sugo, J. Okamoto, Application of adsorption unit charged with amidoxime capillary fibers to recovery of uranium from seawater utilizing flow of ocean current. Nihon Genshiryoku Gakkaishi **32**, 919–924 (1990)
63. K. Saito, T. Yamaguchi, K. Uezu, S. Furusaki, T. Sugo, J. Okamoto, Optimum preparation conditions of amidoxime hollow fiber synthesized by radiation-induced grafting. J. Appl. Polym. Sci. **39**, 2153–2163 (1990)
64. T. Takeda, K. Saito, K. Uezu, S. Furusaki, T. Sugo, J. Okamoto, Adsorption and elution in hollow-fiber-packed bed for recovery of uranium from seawater. Ind. Eng. Chem. Res. **30**, 185–190 (1991)
65. K. Sekiguchi, K. Saito, S. Konishi, S. Furusaki, T. Sugo, H. Nobukawa, Effect of seawater temperature on uranium recovery from seawater using amidoxime adsorbents. Ind. Eng. Chem. Res. **33**, 662–666 (1994)

66. A. Katakai, N. Seko, T. Kawakami, K. Saito, T. Sugo, Adsorption of uranium in sea water using amidoxime adsorbents prepared by radiation-induced cografting. Nihon Genshiryoku Gakkaishi **40**, 878–880 (1998)

67. A. Katakai, N. Seko, T. Kawakami, K. Saito, T. Sugo, Adsorption performance in sea water of amidoxime nonwoven fabrics prepared by radiation-induced cografting of acrylonitrile and methacrylic acid. Bull. Soc. Sea Water Sci. Jpn. **53**, 180–184 (1999)

68. T. Kawai, K. Saito, K. Sugita, T. Kawakami, J. Kanno, A. Katakai, N. Seko, T. Sugo, Preparation of hydrophilic amidoxime fibers by cografting acrylonitrile and methacrylic acid from an optimized monomer composition. Radiat. Phys. Chem. **59**, 405–411 (2000)

69. T. Kawai, K. Saito, K. Sugita, A. Katakai, N. Seko, T. Sugo, J. Kanno, T. Kawakami, Comparison of amidoxime adsorbents prepared by cografting of methacrylic acid and 2-hydroxyethyl methacrylate with acrylonitrile onto polyethylene. Ind. Eng. Chem. Res. **39**, 2910–2915 (2000)

70. N. Seko, A. Katakai, S. Hasegawa, M. Tamada, N. Kasai, H. Takeda, T. Sugo, K. Saito, Aquaculture of uranium in seawater by a fabric-adsorbent submerged system. Nucl. Technol. **144**, 274–278 (2003)

71. D. Kudo, Y. Matuzaki, S. Kawai-Noma, D. Umeno, K. Saito, Preparation of anion-exchange fiber with radiation-induced emulsion graft polymerization for rapid protein purification. Radioisotopes **66**, 1–7 (2017)

72. N. Seko, L.T. Bang, M. Tamada, Syntheses of amine-type adsorbents with emulsion graft polymerization of glycidyl methacrylate. Nucl. Instrum. Methods Phys. Res. B **265**, 146–149 (2007)

73. N. Seko, N.T.Y. Ninh, M. Tamada, Emulsion grafting of glycidyl methacrylate onto polyethylene fiber. Radiat. Phys. Chem. **79**, 22–26 (2010)

74. S. Sugiyama, S. Tsuneda, K. Saito, S. Furusaki, T. Sugo, K. Makuuchi, Attachment of sulfonic acid groups to various shapes of polyethylene, polypropylene and polytetrafluoroethylene by radiation-induced graft polymerization. React. Polym. **21**, 187–191 (1993)

75. S. Tsuneda, K. Saito, S. Furusaki, T. Sugo, K. Makuuchi, Simple introduction of sulfonic acid group onto polyethylene by radiation-induced cografting of sodium styrenesulfonate with hydrophilic monomers. Ind. Eng. Chem. Res. **32**, 1464–1470 (1993)

76. S. Tsuneda, K. Saito, H. Mitsuhara, T. Sugo, Novel ion-exchange membranes for electrodialysis prepared by radiation-induced graft polymerization. J. Electrochem. Soc. **142**, 3659–3663 (1995)

77. K. Miyoshi, T. Miyazawa, N. Sato, D. Umeno, K. Saito, T. Nagatani, N. Yoshikawa, Development of novel ion-exchange membranes for electrodialysis of seawater by electron-beam-induced graft polymerization (I) Selection of trunk polymeric films. Bull. Sea Water Sci. Jpn. **63**, 167–174 (2009)

78. T. Miyazawa, Y. Asari, K. Miyoshi, D. Umeno, K. Saito, T. Nagatani, N. Yoshikawa, Development of novel ion-exchange membranes for electrodialysis of seawater by electron-beam-induced graft polymerization (II) Graft polymerization of vinyl benzyltrimethylammonium chloride and sodium styrenesulfonate onto nylon-6 film. Bull. Soc. Sea Water Sci. Jpn. **63**, 175–183 (2009)

79. Y. Asari, T. Miyazawa, K. Miyoshi, D. Umeno, K. Saito, T. Nagatani, N, Yoshikawa, Development of novel ion-exchange membranes for electrodialysis of seawater prepared by electron-beam-induced graft polymerization (lll) Co-graft polymerization of glycidyl methacrylate and divinylbenzene onto high-density polyethylene film. Bull. Soc. Sea Water Sci. Jpn. **63**, 387–394 (2009)

80. T. Miyazawa, Y. Asari, K. Miyoshi, D. Umeno, K. Saito, T. Nagatani, N. Yoshikawa, R. Motokawa, S. Koizumi, Development of novel ion-exchange membranes for electrodialysis of seawater by electron-beam-induced graft polymerization (lV) polymeric structures of cation-exchange membranes based on nylon-6 film. Bull. Soc. Sea Water Sci. Jpn. **64**, 360–365 (2010)

81. K. Ishimori, T. Miyazawa, Y. Asari, D. Miyoshi, D. Umeno, K. Saito, K. Mizuguchi, T. Aritomo, K. Yoshie, Preparation of mono-valent cation selective cation-exchange membranes for electrodialysis of seawater by electron-beam-induced graft polymerization. Bull. Soc. Sea Water Sci. Jpn. **65**, 35–41 (2011)

82. N. Yoshikawa, Report of research institute of salt and sea water science. Salt Ind. Cent. Jpn. **10**, 25–28 (2008)

83. K. Miyazaki, N. Shoji, Y. Asari, K. Miyoshi, D. Umeno, K. Saito, Preparation of heat- and alkali-resistant anion-exchange membranes by electron-beam-induced graft polymerization of bromo-butyl styrene onto polyethylene film. Membrane (Maku) **35**, 305–310 (2010)

84. Y. Asai, N. Shoji, K. Miyoshi, D. Umeno, K. Saito, Electrodialysis of sulfuric acid with cation-exchange membranes prepared by electron-beam-induced graft polymerization. J. Ion Exch. **22**, 53–57 (2011)

85. Y. Sekiya, Y. Shimoda, D. Umeno, K. Saito, G. Furumoto, H. Shirataki, N. Shinohara, N. Kubota, Preparation of cation-exchange particle designed for high-speed collection of proteins by radiation-induced graft polymerization. J. Ion Exch. **21**, 29–34 (2010)

86. T. Harayama, Y. Okamura, Y. Shimoda, D. Umeno, K. Saito, N. Shinohara, N. Kubota, Protein resolution in elution chromatography using novel cation-exchange polymer-brush-immobilized particles. J. Chem. Eng. Jpn. **45**, 896–902 (2012)

87. T. Someya, Y. Okamura, G. Wada, Y. Shimoda, D. Umeno, K. Saito, N. Shinohara, N. Kubota, Comparison of resolution of proteins in elution chromatography between cation-exchange polymer brush immobilized particle- and commercially available cation-exchange-bead-packed beds. J. Ion Exch. **24**, 1–7 (2013)

88. Y. Shimoda, Y. Sekiya, D. Umeno, K. Saito, G. Furumoto, H. Shirataki, N. Shinohara, N. Kubota, Protein-binding characteristics of anion-exchange particles prepared by radiation-induced graft polymerization at low temperatures. J. Chem. Eng. Jpn. **46**, 588–592 (2013)

Chapter 5
Competition Between Graft Chains and Rivals

Abstract New polymeric adsorbents prepared by radiation-induced graft poly-merization are superior to conventional adsorbents in terms of resolution in elution chromatography and dynamic binding capacity in the flow-through mode. However, currently used adsorbents cannot be easily replaced with our new graft-type adsorbents. Thus far, graft-type adsorbents have not been considered as alternatives. This chapter provides information indicating that graft-type adsorbents may be useful for separation. When a new need for separation that cannot be easily met using existing adsorbents arises, then graft-type adsorbents will be available as promising candidates to meet this need.

Keywords Resolution · Elution chromatography · Flow-through mode
Graft-type adsorbent

We have argued that polymeric adsorbents prepared by radiation-induced graft poly-merization have revolutionized separations and reactions involving ions and mole-cules. We must demonstrate the advantages of graft-type polymeric adsorbents over conventional ones. Our polymeric adsorbents exhibit higher performance in separation because graft chains are appended to different quantities and structures of trunk polymers and because the extended polymer brushes can bind proteins in multilayers and extractants at high densities. In this chapter, we describe the results of comparison of separation performance between our adsorbents and commercialized adsorbents.

5.1 Graft-Particle-Packed Bed Versus Commercialized Bead-Packed Bed

5.1.1 Elution Chromatography of Proteins [1]

Adsorbents capable of purifying proteins have been manufactured by Pharmacia Co. in Sweden, which merged with GE Healthcare Co. in 2004. Different resin beads based on agarose as a natural polymer are now commercially available for affinity, ion-exchange, and hydrophobic interaction chromatography. In addition, technical notes on these beads included descriptions of their extensive applications to biochemistry and biotechnology with references. We compared the resolution of three proteins as models in elution chromatography between a column charged with ion-exchange polymer-brush-immobilized particles prepared by radiation-induced graft polymerization and one charged with ion-exchange beads purchased from GE Healthcare Co.

Protein purification by elution chromatography has already been extensively applied in the pharmaceutical industry. Many types of ion-exchange bead with columns are commercially available. A type of cation-exchange bead produced by GE Healthcare Co., SOURCE 30S, was selected for comparison; we prepared cation-exchange particles by radiation-induced graft polymerization of glycidyl methacrylate onto polyethylene particles and subsequently conversion of the epoxy group into a sulfonic acid group ($-SO_3H$), as shown in Fig. 2.10. The resultant cation-exchange particles are referred to as SS particles.

We then compared the performance between the top-ranking cation exchangers SOURCE 30S beads and the new SS particles. The properties of these two cation exchangers are listed in Table 5.1. The SS particles are not uniform in size, with an average size of 60 μm, whereas the SOURCE 30S beads have a relatively uniform size of 30 μm. When the particles and beads were packed into columns with an inner diameter of 5 mm to identical heights, the SS-particle-packed column (SS column) exhibited a higher pressure required for the liquid to flow through it than the SOURCE 30S-packed column (SOURCE column) at the same flow rate. In order to ensure an unbiased comparison, the heights of the packed SS particles and SOURCE 30S beads were set at 1.5 and 2.5 cm, respectively, enabling the same pressure loss in the columns and hence the same resistance to flow.

Three proteins as models, i.e., chymotrypsinogen (Chy), cytochrome C (Cyt), and lysozyme (Lys), were loaded on the top of columns at a mixing ratio of Chy: Cyt: Lys = 3: 4: 3, in accordance with instruction in a catalogue from GE Healthcare Co. By gradient elution, i.e., raising the sodium chloride concentration in a buffer (20 mmol/L phosphate buffer, pH 6.0), the three proteins were eluted in the order of the strength of their interactions with the eluent flowing downward through the column. This method is referred to as gradient elution chromatography.

Elution chromatograms or protein concentration profiles at the column exit during elution were determined for various amounts of loaded proteins and flow rates of the eluent. The higher the amount of loaded proteins and the higher the flow rate of the eluent, the higher the throughput of protein purification by elution chromatography. Elution chromatograms of the SS and SOURCE columns for various amounts of loaded proteins at an eluent velocity of 300 cm/h are shown in Fig. 5.1a, b, respectively. The retention volumes of the peaks indicated that the proteins were separated and the overlapping of the tails was quantitatively small. The separation factors for each column for Chy/Cyt and Cyt/Lys are shown in Fig. 5.2a, b, respectively. For both combinations, the SS column was superior to the

Table 5.1 Properties of cation exchangers for protein separation

	SS particle prepared by Chiba University	SOURCE 30S manufactured by GE Healthcare
Matrix	Polyethylene	Styrene-DVB copolymer
Size	Particle >60 μm	Bead 30 μm
SO$_3$H group density (mmol/mL-bed)	0.27	0.30
Equilibrium binding capacity for lysozyme (mg/mL-bed)	28	84
Pressure loss (MPa)	0.05	0.03

SOURCE column. Our graft-type particles showed better separation than the representative commercially available beads.

During loading, proteins reach the upper interface of the polymer brushes appended to the surface of polyethylene particles and diffuse into the depth along the graft phase driven by the gradient of the amount of adsorbed proteins. During elution, the proteins diffuse toward the bulk of the eluent flowing through the interstices between the particles. There are two reasons for the superior performance of the SS particles: (1) a shorter diffusional mass-transfer path of the proteins from the ion-exchange groups of polymer brushes and (2) a stronger interaction of the proteins with the polymer brushes. First, the length of the diffusional mass-transfer path is comparable to that of polymer brushes, which is much smaller than the radius of SOURCE 30S beads. Second, the extended cation-exchange polymer

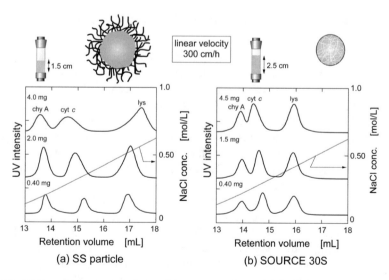

Fig. 5.1 Elution chromatograms for varying amounts of proteins loaded onto SS-particle- and SOURCE 30S-packed beds

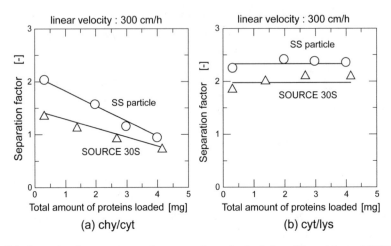

Fig. 5.2 Separation factor versus total amount of proteins loaded on SS-particle- and SOURCE 30S-packed beds. Reprinted with permission from Ref. [1]. Copyright 2013 Japan Society of Ion Exchange

brushes can hold proteins in multilayers via multipoint binding. The multipoint binding of proteins strengthens the electrostatic interaction of the proteins with the cation-exchange polymer brushes. This strengthened electrostatic interaction allows distances between the peaks in the elution chromatogram to widen and the tails of the peaks to narrow in chromatograms.

5.1.2   Elution Chromatography of Nd and Dy Ions [2, 3]

In the purification of the three proteins by elution chromatography, the graft-type ion exchanger was superior to a conventional ion exchanger. The merit of this polymeric adsorbent compared with commercially available adsorbents is that it is applicable to the extraction of rare-earth metals (REMs) when impregnated with extractants. LEWATIT or HDEHP-impregnated beads, manufactured by LANXESS Co., are commercially available adsorbent. We immobilized bis(2-ethylhexyl) phosphate (HDEHP) as an acidic extractant onto the hydrophobic moiety of polymer chains grafted onto polyethylene particles. The resultant HDEHP-impregnated particles are referred to as CHIBATIT.

Two rare-earth metals, Nd and Dy, were extracted from a fluid containing cutting powder from neodymium magnets. The preparation scheme for the polymeric adsorbent consists of three steps: First, glycidyl methacrylate (GMA) is graft-polymerized, with a degree of grafting of 50%, onto electron-beam-irradiated polyethylene particles with an average size of 35 μm. Second, the epoxy groups of the graft chains are converted into dodecylamino groups ($-NHC_{12}H_{25}$). Third, by

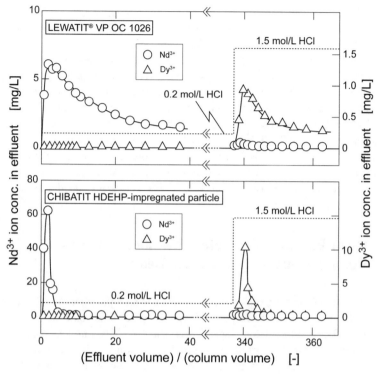

Fig. 5.3 Elution chromatograms of Nd^{3+} and Dy^{3+} ions in CHIBATIT prepared by Chiba Univ. and LEWATIT® manufactured by LANXESS

immersing the hydrophobic-ligand-immobilized particles in HDEHP solution, 0.25 mmol/g HEDHP is impregnated into the particles.

The properties of CHIBATIT and LEWATIT are listed in Table 5.2. For CHIBATIT, HDEHP was impregnated into the graft chains extending from the surface of the polyethylene particles. For LEWATIT, most of the HDEHP was immobilized into the entire interior surface of the porous polymeric beads; therefore, the amount of HDEHP impregnated into LEWATIT was threefold that into

Table 5.2 Properties of HDEHP-impregnated particles and beads for rare metal purification

	HDEHP-impregnated particles prepared by Chiba University CHIBATIT	HDEHP-impregnated beads manufactured by LANXESS LEWATIT®
Form of support	Particle (>35 μm)	Bead (300–1600 μm)
Amount of HDEHP impregnated (mmol/g)	0.25	0.91
Site of HDEHP impregnation	Near the particle surface	Throughout the bead

CHIBATIT. However, because the diffusional mass-transfer path of CHIBATIT for rare-earth metal ions is much shorter than that of LEWATIT, high-performance elution chromatography using CHIBATIT is possible.

CHIBATIT and LEWATIT were each packed into a column with an inner diameter of 5 mm to a height of 2 cm. Both Nd and Dy ions were loaded on the top of the column. Subsequently, Nd and Dy ions were separated by a stepwise increase in hydrochloric acid (HCl) concentration in the eluent. Two concentrations of HCl, i.e., 0.2 and 1.5 M HCl, were used to quantitatively elute Nd and Dy ions, respectively. Elution chromatograms of CHIBATIT and LEWATIT are shown in Fig. 5.3 for identical amounts of Nd and Dy ions loaded and identical flow rates of HCl as the eluent. The first and second peaks of CHIBATIT corresponding to Nd and Dy ions, respectively, were narrower than those of LEWATIT; CHIBATIT was clearly superior to LEWATIT.

5.2   Graft Porous Hollow-Fiber Membrane Versus Commercialized Bead-Packed Bed [4]

To purify a target protein, e.g., antibodies, for use as a pharmaceutical compound, undesirable proteins must be removed from biological fluids such as broth, serum, and milk. Here, ion-exchange adsorbents capable of removing proteins were compared in the flow-through mode. A protein solution was forced to flow through a bead-packed column, and the protein concentration at the column exit was determined as a function of accumulated effluent volume. The resultant profile was referred to as a breakthrough curve. The breakthrough point is defined as the accumulated effluent volume when the effluent concentration reaches 5 or 10% of the feed concentration. For practical use, the adsorption process is stopped before undesirable proteins are detected in the effluent.

Commercially available anion-exchange beads of two types were packed into a column, the volume of which was almost identical to that of an anion-exchange porous hollow-fiber membrane, with inner and outer diameters of 2.8 and 4.4 mm, respectively, and a length of 72 mm. One was of the gel type, DEAE Sepharose™ FF, manufactured by GE Healthcare Co. The other was of the perfusion bead type, POROS™ 50HQ, manufactured by PerSeptive Biosystems. The perfusion beads have two sizes of pores, i.e., a "throughpore" and micropores surrounding the throughpore. Proteins are transported by convective flow of a protein solution through the throughpore into the micropore and are captured by the anion-exchange groups introduced into the polymer chains forming the micropores. The perfusion beads minimize diffusional mass-transfer resistance owing to the convective flow of the protein solution.

The properties of three anion exchangers are summarized in Table 5.3. Albumin solution as a model buffered with 20 mMTris-HCl buffer (pH 8.0) was forced to permeate across the hollow fibers or flow through the bead-packed columns. Negatively charged albumin (pI4.9) at pH 8.0 is adsorbed to the anion-exchange groups. Breakthrough curves, i.e., the protein concentration change as a function of effluent volume at the outside surface of the hollow fiber or the exit of the column, were determined at various flow rates of the protein solutions. Dynamic binding capacity per volume of hollow fibers or columns, where the volume of the hollow fibers includes the volume of the lumen, was evaluated from the breakthrough curves according to Eq. (3.1). The breakthrough point was defined as the effluent volume where the effluent concentration reached 10% of the feed concentration. The operating pressure required for the solution to permeate across the hollow fiber or flow through the column was also measured.

A high dynamic binding capacity and a low operating pressure are desirable for protein purification. The space velocities (SVs) of the hollow fibers and two columns are shown in Fig. 5.4 as a function of operating pressure. SV is defined by dividing flow rate by the volume of hollow fibers including the lumen or columns. The hollow fibers exhibited a lower flow resistance than the columns. The dynamic binding capacities are shown in Fig. 5.5 as a function of SV. The SV range of $92–1020 \ h^{-1}$ is converted into the average residence time of the solution: 39–3.5 s. The dynamic binding capacity of DEA-EA fibers remained constant irrespective of SV because the diffusional mass-transfer resistance of proteins to the anion-exchange group of the graft chains in the pores is negligible. This feature of the modified porous hollow-fiber membrane is effective only when the anion-exchange interaction is instantaneous.

In contrast, the dynamic binding capacities of the two columns decreased with increasing SV. This can be explained by noting that the time required for proteins to diffuse into the beads is reduced as flow rate increases. The DEA-EA fibers are superior to the bead-packed columns in that a high-throughput recovery of proteins is attainable.

Moreover, the DEA-EA fibers possess the additional merit of linear scale-up. For a large-scale recovery of proteins using a bead-packed column, the determination of the column height is necessary because column diameter is restricted: A versatile

Table 5.3 Properties of anion-exchange hollow fibers and beads for protein purification

	DEA-EA hollow-fiber membrane prepared by Chiba University	QA-P bead manufactured by PerSeptive Biosystems	DEAE-G bead manufactured by Pharmacia Biotech
Matrix	Polyethylene	Styrene-DVB copolymer	Crosslinked agarose
Anion-exchange group	Diethylamino group	Quaternary ammonium salt	Diethylaminoethyl group
Anion-exchange-group density	1.7 mmol/g		0.11–0.16 mmol/mL
Form	Porous hollow-fiber membrane	Bead	Bead
Size	Inner diameter: 2.8 mm Outer diameter: 4.4 mm	50 μm	90 μm

Fig. 5.4 Space velocity as a
function of operating pressure
for two types of
anion-exchange porous
hollow-fiber membrane (EA
and DEA-EA membranes)
and two types of
commercially available
anion-exchange bead
(DEAE-G and QA-P beads).
Reprinted with permission
from Ref. [4]. Copyright 1996
John Wiley & Sons Ltd

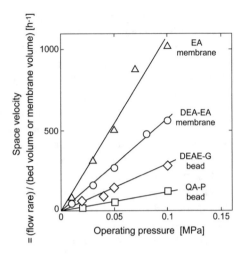

Fig. 5.5 Dependence of
dynamic binding capacity on
space velocity for two types
of anion-exchange porous
hollow-fiber membranes and
two types of commercially
available anion-exchange
beads. Reprinted with
permission from Ref. [4].
Copyright 1996 John Wiley &
Sons Ltd

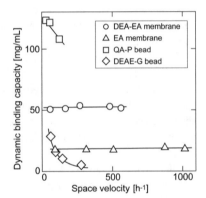

dimension ratio of diffusional mass-transfer path length to flow-through length
requires a nonlinear scale-up. In contrast, the scale-up of the hollow fibers is
achieved by increasing the number of hollow fibers because the length of the
diffusional mass-transfer path or membrane thickness remains constant.

5.3   Graft Porous Sheet Versus Commercialized Bead-Packed Bed [5]

Solid-phase extraction (SPE) is a versatile and powerful method for the clean-up or
preconcentration of molecules or ions in analytical applications. To date, the liquid–
liquid extraction technique has been adopted to enrich a target substance on the
basis of the difference in its solubility in water and organic phases. For example,
chloroform or benzene may be adopted as the organic phase. In place of the organic

phase, a polymer and silica gel of a bead containing functional groups capable of specifically capturing the target substance are used. Because these solid phases are regarded as solid adsorbents, the SPE technique may be called an adsorption method.

When a target ion in a sample solution is required to be enriched 1000-fold whereas foreign ions are removed, 1 L of a sample solution flows through an adsorbent-packed column and target ions are bound to the adsorbent. Subsequently, the target ions are quantitatively eluted with 1 mL of an eluent such as an acid or methanol. Through a series of steps, i.e., adsorption, washing, and elution, the target ions are enriched 1000-fold.

Products of SPE obtained on the basis of various interaction modes between the analyte and the adsorbent have been commercialized by many companies. The interaction modes include ion-exchange, chelate-formation, and hydrophobic interactions. We selected Empore™, manufactured by 3M Co., for comparison.

Atomic absorption spectroscopy (AAS) and induced-coupled plasma (ICP)-AES or ICP–MS have been used to directly determine metal ions in sample solutions. However, when the target ions dissolve at low concentrations or compete with various foreign metal ions, the sample should be enriched before introducing it into analytical instruments. This clean-up or preconcentration raises the precision and reproducibility of the analysis.

We set up a competition involving SPE designed for the enrichment of heavy metal ions. The rival for our chelating porous sheet was the Empore™ chelate cartridge manufactured by 3M Co. The chelate-forming group common to both adsorbents is the iminodiacetate group.

The porous structure of our graft-type porous sheet is illustrated in Fig. 5.6. Our graft-type adsorbent captures heavy metal ions through the polymer chains grafted onto polyethylene porous sheets. The polymer brushes extending from the pore surface toward the pore interior wait for the heavy metal ions to diffuse from the bulk flowing through the pore. In contrast, the chelating cartridge contains relatively small chelating beads entangled in Teflon fibers. The smaller the beads, the shorter the diffusion time of the heavy metal ions into the beads. Teflon fibers provide the interstices through which the sample solution flows with an appropriate pressure loss. This hybrid structure sacrifices the binding capacity of heavy metal ions; however, the clean-up for analytical use may not require a high binding capacity of adsorbents.

Breakthrough curves for the permeation or flow through of a copper chloride solution are compared in Fig. 5.7a between the chelating porous sheet- and chelating bead-packed cartridges. For this comparison, our cartridge was fabricated by cutting our chelating porous sheet to pack it into a commercially available empty cartridge designed for SPE.

The breakthrough curves of the chelating porous sheet for copper ions overlapped irrespective of the residence time in the range from 3 to 33 s. Namely, an identical amount of copper ions was captured by the cartridge from an identical volume of the sample solution irrespective of the flow rate of the solution. This

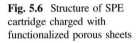

Fig. 5.6 Structure of SPE
cartridge charged with
functionalized porous sheets

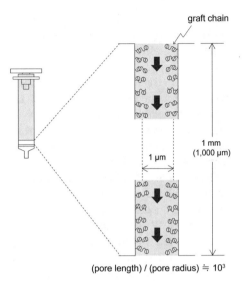

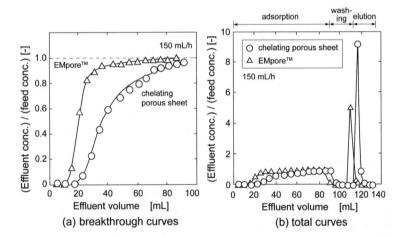

Fig. 5.7 Breakthrough, washing, and elution curves for chelating porous sheet-packed cartridge
prepared by Chiba University versus Empore™ chelating cartridge manufactured by 3M Co.
Reprinted with permission from Ref. [5]. Copyright 2007 Elsevier

characteristic is favorable for users because the users do not have to carefully
control flow rate.

The properties of both cartridges are listed in Table 5.4. The adsorption and
elution performances are shown in Fig. 5.7b. Our graft-type cartridge exhibited
higher binding capacity for copper ions than the Empore™ chelate cartridge
because of its higher chelating group density. A sharper peak for our graft-type
cartridge corresponding to copper ions in the elution curve than for the Empore™

chelate cartridge was observed because the diffusional mass-transfer resistance was minimized. Our graft-type cartridge performed better than the Empore™ chelate cartridge.

5.4 Graft-Fiber-Packed Bed Versus Commercialized Bead-Packed Bed [6, 7]

5.4.1 Removal of Boron

Fibers as adsorbents designed for separation and reaction have the following advantages over beads: (1) a shorter diffusional mass-transfer path and a larger external surface area and (2) versatile fabrication into various forms such as braids and wound filters. However, to date, a method of modifying existing fibers into fibers useful for separation and reactions has not been developed.

The external surface area of a bead with the diameter d can be calculated as

$$\pi d^2 / [(1/6)\pi d^3] = 6/d, \tag{5.1}$$

whereas the external surface area of a long fiber with the diameter of d can be calculated as

$$\pi d L / [(1/4)\pi d^2 L] = 4/d, \tag{5.2}$$

Thus, for an identical diameter, a bead has 1.5-fold higher external surface area than a fiber. However, a fiber is regarded as a fusion of small beads linked together. Adsorbent fibers are more advantageous than adsorbent beads because they are easier to handle.

An adsorbent fiber can be packed into a column at various packing densities. A low packing density may produce a distribution inflow resistance, resulting in the channeling of liquid flow. On the other hand, a high packing density may increase the total density of functional groups per column. In contrast, beads are very closely packed unless section plates are placed into the column. The column is often tapped

Table 5.4 Properties of chelating porous sheet- and particle-packed cartridge

	IDA porous sheet prepared by Chiba University	Empore™ chelate cartridge manufactured by 3M
Matrix	Polyethylene	Styrene-DVB copolymer
Volume (mL)	0.36	0.07
Weight (g)	0.12	0.06
Pure water flux (m/h)[a]	12	9.0

[a]0.1 MPa, 25 °C

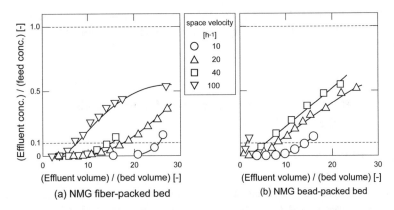

Fig. 5.8 Breakthrough curves as a function of space velocity for NMG fibers prepared by Chiba University versus commercially available NMG beads manufactured by Mitsubishi Chemical Co. Reprinted with permission from Ref. [6]. Copyright 2011 American Chemical Society

after beads have been charged in the column to achieve high reproducibility of packing.

We now compare the removal of boron from liquids between our graft-type chelating fibers and a type of commercially available chelating bead. Both the fibers and the beads contain the N-methylglucamine (NMG) moiety as a chelate-forming group with boron. The chelating fiber is referred to as an NMG fiber. DIAION™ CRB05 beads, manufactured by Mitsubishi Chemical Co., were used for comparison. The properties of the adsorbents and the characteristics of the adsorbent-packed column are listed in Table 5.5.

Table 5.5 Properties of beds packed with N-methylglucamine-immobilized fibers and beads for boron removal

	NMG fiber prepared by Chiba University	NMG bead manufactured by Mitsubishi Chemical DIAION™ CRB05
Adsorbent		
Form	Fiber	Bead
Size (μm)	55 (fiber diameter)	>400 (diameter)
NMG group density (mmol/g)	2.0	2.7
Equilibrium binding capacity for boron (mg-B/g)	12	12
Column		
(Weight of fiber)/(loading volume) (g/mL-bed)	0.38	0.43
Equilibrium binding capacity for boron (mg-B/mL-bed)	2.5	1.0

Fig. 5.9 Dependence of
dynamic binding capacity of
boron on space velocity for
NMG fibers vs NMG beads.
Reprinted with permission
from Ref. [6]. Copyright 2011
American Chemical Society

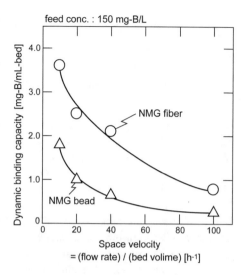

We graft-polymerized glycidyl methacrylate (GMA) onto nylon-6 fibers with
200% degree of grafting. Subsequently, some of the epoxy groups were converted
into NMG moieties at a molar conversion of 80%. As a result, the density of NMG
moieties immobilized onto the fibers was 2.0 mmol/g, which was lower by 26%
than that of the NMG beads (2.7 mmol/g). A higher degree of grafting and higher
molar conversion can produce brittle fibers. The fibers at an NMG moiety density of
2.0 mmol/g possess an acceptable mechanical strength.

Adsorbent fibers and beads were compared when both were packed into the same
kind of column. Breakthrough curves of the NMG-fiber- and NMG-bead-packed
columns for boron are shown in Fig. 5.8a, b, respectively, as a function of SV, where
SV is defined as a ratio of flow rate to the column volume. Dynamic binding
capacities at various SVs were determined from the breakthrough curves (Fig. 5.9).
The NMG-fiber-packed column exhibited a higher dynamic binding capacity than
the NMG-bead-packed column in this SV range: The NMG fibers outperformed the
NMG beads. The shorter diffusional mass-transfer path and larger specific external
surface area of the NMG fibers contribute to this better performance.

5.4.2 Purification of Antibody Drugs [8]

Antibody drugs have gained key positions in the pharmaceutical market because of
their high effectiveness and mild side effects. Currently, antibody drugs are used for
the therapy of uterine and breast cancers. An example of the production process of
an antibody drug is shown in Fig. 5.10. Among these processes, the purification of

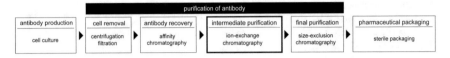

Fig. 5.10 Production process of antibody drug

target antibodies and the removal of impurities using various chromatographic techniques such as affinity and ion-exchange chromatographies can account for 80% of the total production cost of antibody drugs.

The target antibody and small amounts of impurities are contained in the liquid obtained by the elution of proteins bound to affinity beads packed into a column. Then, the impurities are captured in a flow-through mode by ion exchangers packed into a column while minimizing the loss of the target antibody. Therefore, an ion exchanger with high binding capacity for the impurities is necessary to reduce the frequency with which the column must be replaced.

We prepared a weakly acidic cation-exchange fiber by the radiation-induced graft polymerization of acrylic acid (AAc) onto nylon-6 fiber. The dynamic binding capacity (DBC) of the fiber-packed column for impurities such as foreign proteins was compared with that of a bead-packed column. Commercially available weakly acidic cation-exchange beads (CM Sepharose Fast Flow, GE Healthcare) were adopted as the beads.

The properties of the AAc-grafted fiber and CM Sepharose Fast Flow are listed in Table 5.6. Both ion exchangers were packed into a 1.5-cm-high and 0.90-cm-diameter column. Lysozyme solution buffered at pH 6.0 flowed downward through the column. Breakthrough curves, i.e., changes in the effluent concentration of lysozyme at the exit of the column with increasing effluent volume, are depicted in Fig. 5.11. The DBC of the column for lysozyme, which was defined as the amount of lysozyme adsorbed until the effluent concentration reaches 10% of the feed concentration, was evaluated from the breakthrough curve.

Table 5.6 Comparison of properties between AAc-grafted fiber and CM Sepharose fast flow

	AAc-grafted fiber	CM Sepharose FF
Shape	Fiber	Porous beads
Adsorption site	Polymer brush of graft chain	Surface of pores
Matrix	Nylon	Agarose
Total cation-exchange capacity (mmol/mL-column)	1.2	0.09–0.13
Column height (cm)	1.5	1.5
Packing density (g-adsorbent/mL-column)	0.20	–
Saturated adsorption amount of lysozyme (mg/mL-column)	400[a]	190

[a]Calculated by dividing saturated adsorption amount by density of packing

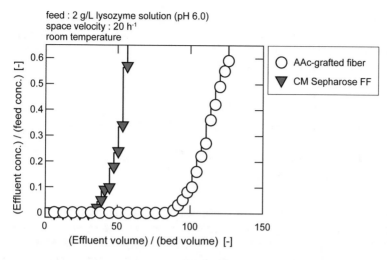

Fig. 5.11 Breakthrough curves of columns charged with AAc-grafted fiber and CM Sepharose Fast Flow for lysozyme

DBCs of the fiber- and bead-packed columns are shown as a function of space velocity (SV) at various buffer concentrations in Fig. 5.12. As SV increased, in other words, as the residence time of the liquid across the column decreased, the DBC of the CM Sepharose-FF-bead-packed column decreased with an increase in buffer concentration. In contrast, the AAc-grafted fiber-packed column exhibited values of DBC independent of the buffer concentration from 50 to 150 mmol/L. This notable feature in the dependence of the DBC of the fiber-packed column on the buffer concentration is because proteins are bound by the negatively charged polymer brush at more points at a lower buffer concentration and because proteins

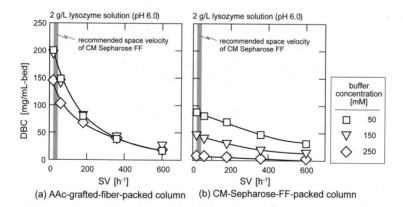

Fig. 5.12 Dynamic binding capacities of fiber- and bead-packed columns as a function of SV at various buffer concentrations

are bound at fewer points at a higher buffer concentration. The operational pressure of the fiber-packed column required to attain an identical SV was one-quarter that of the bead-packed column, which is an additional advantage of the fiber-packed column over the bead-packed column.

5.5 Graft Fiber Versus Commercialized Granules

5.5.1 Removal of Catechin from Green-Tea Extractant [9, 10]

The annual amount of green tea consumed as a beverage has been increasing in Asian countries including Japan. Catechin is extracted from green-tea leaves by their immersion in cold or hot water. Catechin, categorized as polyphenols, has a bitter taste and stimulating effect; therefore, the adjustment of the catechin content in green-tea drinks is desirable for children and the elderly.

The catechin contents in some green-tea drinks can be adjusted by adding commercially available PVPP powder, i.e., polyvinyl N-polypyrrolidone powder, to the green-tea extractant. PVPP powder does not adsorb other components, which would change the flavor of the drink. After the removal of catechin from the extractant liquid, the PVPP powder is removed by filtering and discarded without any recycling. Instead of PVPP powder, recyclable adsorbents capable of removing catechin are required from an environmental viewpoint.

We prepared a recyclable adsorbent in a fiber form by the radiation-induced graft polymerization of N-vinyl-2-pyrrolidone (NVP) onto nylon-6 fiber. As illustrated in Fig. 5.13, catechin is bound to the amide moiety of NVP through hydrogen bonding. N-vinylacetamide (NVAA) as a grafted vinyl monomer is a promising candidate for interacting with catechin because an NVAA molecule has a less distorted structure than an NVP molecule. The preparation schemes for NVP- and NVAA-grafted fibers are shown in Fig. 5.14. A previously irradiated nylon-6 fiber was immersed in an aqueous solution of NVP or NVAA to form the graft chain.

The properties of three different adsorbents for catechin are listed in Table 5.7. The density (5.2 mmol/g) of the amide moiety of the NVAA fiber was 58% of that (9.0 mmol/g) of PVPP powder. Adsorption isotherms of the adsorbents in aqueous catechin solution at 25 °C are shown in Fig. 5.15. The data were well fitted to the Langmuir equation:

$$q = q_m KC/(1 + KC), \tag{5.3}$$

The obtained values of the saturation capacity q_m and adsorption equilibrium constant K are summarized in Table 5.7. The adsorption equilibrium constant of the PVPP powder (0.0041 L/mg) was similar to that of the NVP-grafted fiber (0.0047 L/mg). The saturation capacity of the NVAA fiber (320 mg/g) was 28% higher than that of the PVPP powder (250 mg/g).

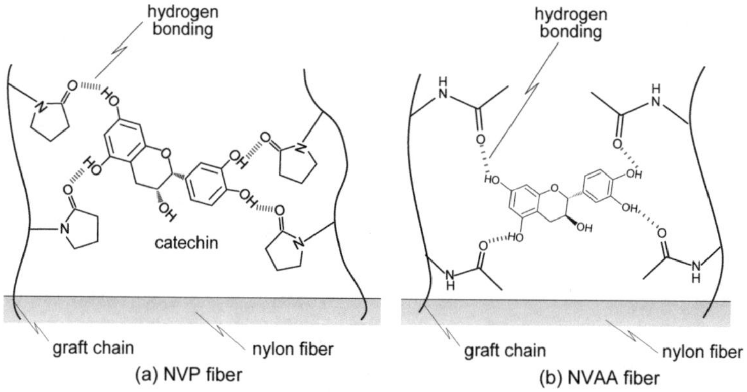

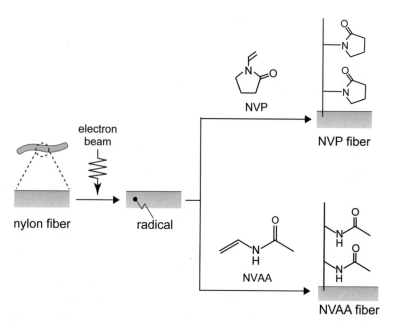

Fig. 5.13 Binding of catechin to amide moieties such as NVP and NVAA

Fig. 5.14 Preparation scheme for NVP- and NVAA-grafted fibers

The catechin adsorbed onto the NVAA fiber was quantitatively eluted with 0.1 M NaOH at 25 °C. Subsequently, the fiber was regenerated by washing with water to adsorb catechin again. During four cycles of adsorption, elution, and washing, the amount of adsorbed catechin remained constant; therefore, the NVAA-grafted fiber is superior to the currently used PVPP powder in that the fiber is readily separated from the green-tea extractant and is recyclable via elution of the catechin with an alkali.

Table 5.7 Properties and Langmuir parameters of various adsorbents for catechin

		NVAA fiber	NVP fiber	PVPP powder
Saturation capacity	(mg-catechin/g)	320	88	250
	(mg-catechin/mmol)	62	18	28
Binding molar ratio (−)		0.15	0.044	0.068
Adsorption equilibrium constant (L/mg)		0.00083	0.0047	0.0041

Fig. 5.15 Adsorption isotherms of various adsorbents for catechin

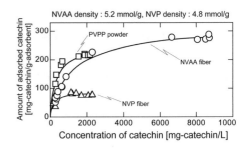

5.5.2 Removal of Radioactive Sr Ions from Contaminated Water [11]

Groundwater is continuously flowing down from the hills toward the reactor buildings of TEPCO's Fukushima Daiichi Nuclear Power Plant (NPP). The amount of groundwater flowing into the NPP has been reduced from 400 to 100 metric tons per day as a result of various countermeasures such as paving the surface around the plant, pumping up groundwater before it reaches the reactor buildings, and installing an impermeable frozen-earth wall. The contact of groundwater with the meltdown nuclear fuel results in the generation of water contaminated with radionuclides such as radioactive Cs and Sr ions. Part of the contaminated water has leaked into the seawater intake area in front of Reactors 1 to 4. The radioactivities of Cs and Sr in surface water must be reduced to below 90 and 30 Bq/L by law, which are approximately 2.8×10^{-8} mg-Cs/L and 5.9×10^{-9} mg-Sr/L, respectively.

The removal of radioactive Cs ions from contaminated water at TEPCO's Fukushima Daiichi NPP was described in Chap. 4. The removal of radioactive Sr ions from contaminated seawater is extremely difficult because nonradioactive Sr ions already exist in seawater at a concentration of 7–8 mg-Sr/L. In addition, Ca and Mg ions, which are also alkaline-earth metal ions similar to Sr ions, have concentrations in seawater of 400 and 1400 mg/L, respectively. Even nonradioactive Sr ions strongly compete with radioactive Sr ions for adsorption sites. Because an adsorbent cannot remove radioactive Sr ions from nonradioactive Sr ions, an essential requirement of adsorbents is high selectivity of Sr compared with Ca and Mg, leading to a high adsorption capacity for Sr.

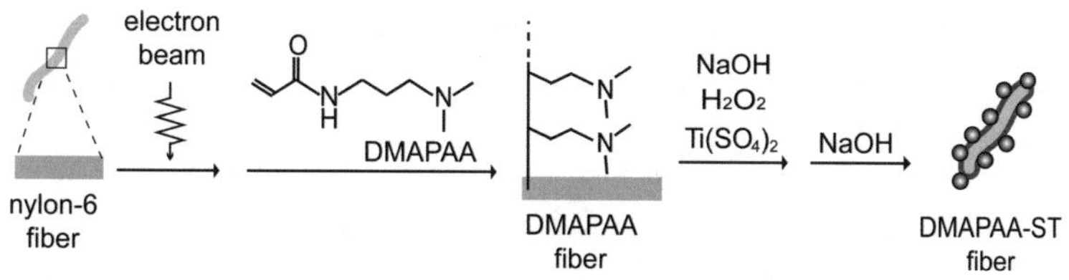

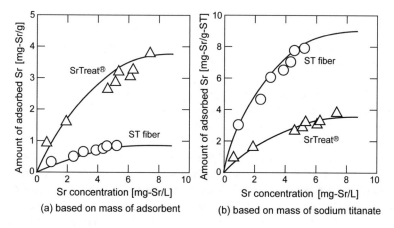

Fig. 5.16 Preparation scheme for sodium titanate (ST)-impregnated fiber

Fortum Co. in Finland has commercialized a Sr-selective adsorbent in a granular form, SrTreat. SrTreat consisting of sodium titanate captures Sr ions selectively via the intercalation of Sr ions between TiO_2 layers. SrTreat granules, with an average size of 100 μm, are relatively brittle and must be packed into a column for practical applications.

We impregnated sodium titanate as a precipitate onto a fiber by radiation-induced graft polymerization and subsequent chemical modification. The preparation scheme of the sodium titanate-impregnated fiber (ST fiber) consists of three steps, as shown in Fig. 5.16. First, dimethylaminopropyl acrylamide (DMAPAA), a vinyl monomer containing a weakly basic anion-exchange group, was graft-polymerized onto a previously irradiated nylon-6 fiber. Second, peroxotitanium complex (POTC) anions, which was prepared by mixing $Ti(SO_4)_2$, H_2O_2, and NaOH, were immobilized as a sodium titanate precursor. Third, sodium hydroxide was fed to hydrolyze the POTC anions in sodium titanate as a precipitate. The precipitate that forms in the graft chain becomes physically entangled with the graft chain. A quantitative dissolution of the precipitate with 1 mol/L nitric acid revealed that the composition ($Na_{3.8}Ti_5O_{11.9}$) of the precipitate on the ST fiber was similar to that of sodium titanate ($Na_4Ti_5O_{12}$) prepared by a hydrothermal reaction [12].

Fig. 5.17 Adsorption isotherms of ST fiber and SrTreat®

Table 5.8 Comparison of Sr removal performance in seawater between ST fiber and SrTreat[®]

Adsorbent	Percentage impregnation	Saturation capacity, q_m		Adsorption equilibrium constant, K
	(%)	(mg-Sr/ g-adsorbent)	(mg-Sr/ g-ST)	(L/mg-Sr)
ST fiber	11	0.97	8.8	0.25
SrTreat[®]	100	3.4	3.4	0.26

The properties of the ST fiber and SrTreat are shown in Table 5.8. The adsorption isotherms of the adsorbents for Sr ions in seawater are shown in Fig. 5.17. The adsorption isotherms for multicomponent adsorption in seawater were approximated by the Langmuir equation; the obtained parameters, i.e., q_m and K, are listed in Table 5.8. Although the value of q_m, i.e., the saturation capacity per gram (0.97 mg-Sr/g) of the ST fiber, was 30% of that (3.4 mg-Sr/g) of SrTreat, the value per gram of sodium titanate immobilized on the ST fiber (8.8 mg-Sr/g) was 2.6-fold higher than that of SrTreat. This indicates that sodium titanate precipitated in the graft chain is effective for the uptake of Sr ions, whereas the sodium titanate constituting SrTreat may be degraded during the granulation.

The adsorption constants were similar for the ST fiber and SrTreat. The ST fiber is advantageous over SrTreat in that the fiber is assembled into braids or a wound filter in consideration of the situation and scale of the decontamination sites where radionuclides are removed from water. In addition, after binding Sr ions, the fiber is burned except for its inorganic portion to reduce its volume for long-term storage as radioactive waste in a high-integrity container (HIC).

5.6 Graft Ion-Exchange Membranes Versus Commercialized Ion-Exchange Membranes [13, 14]

In Japan, edible salt is produced from seawater by electrodialysis and subsequent vacuum drying. Japan consumes electrical power to acquire high-purity edible salt. In contrast, Europe and South America depend on rock salt and lake salt, respectively.

An electrodialyzer consists of approximately 1000 pairs of 2 m^2 cation- and anion-exchange membranes and two electrodes that sandwich them. The electrodialyzer increases the concentration of sodium chloride in solution from 0.5 to 4 M. Electrodialysis is followed by vacuum drying to crystallize sodium chloride.

Salt manufacturing by electrodialysis started in 1960. Because concentrating sodium chloride in a concentration chamber and the electrical resistance of ion-exchange membranes installed in the electrodialyzer is directly related to production cost, manufacturers of ion-exchange membranes have focused their efforts on improving the membranes. For five years after 2005, The Salt Industry Center of

Japan worked on developing novel high-performance ion-exchange membranes by radiation-induced graft polymerization.

Success or failure in the development of novel ion-exchange membranes for electrodialysis depends on the combination of the quality of the trunk polymer and the type of the vinyl monomer used. Nine commercially available films were selected. Also, two vinyl monomers were employed: an originally ion-exchange-group-containing vinyl monomer and a precursor vinyl monomer into which an ion-exchange group could be introduced.

It took two years to develop the ion-exchange membranes, the properties of which were comparable to currently used ion-exchange membranes for electrodialysis, using radiation-induced graft polymerization. The preparation schemes for the ion-exchange membranes that survived the screening for various qualities of the trunk polymer, doses, solvents for vinyl monomer, and grafting temperatures are shown in Fig. 5.18. Their performance in electrodialysis is shown in Fig. 5.19.

From this screening, we learned that a preirradiated nylon-6-made trunk polymer was easily modified with aqueous vinyl monomers in aqueous medium. Because the shape of the film enabled measurements by X-ray diffraction (XRD) and small-angle neutron scattering (SANS), it was possible to observe any changes in the polymeric structure of polyethylene films caused by radiation-induced graft polymerization and subsequent chemical modifications.

Yoshikawa [14] of The Salt Industry Center of Japan succeeded in preparing ion-exchange membranes based on ultrahigh molecular weight polyethylene (UHMWPE) films. The cation- and anion-exchange membranes made from electron-beam-irradiated UHMWPE films exhibited a much better performance than our graft-type cation- and anion-exchange membranes. As shown in Fig. 5.20, UHMWPE-based ion-exchange membranes exhibited excellent performance

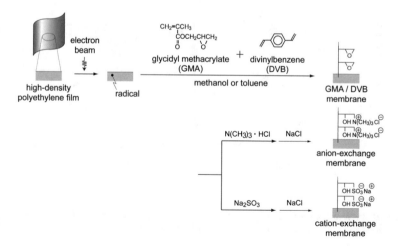

Fig. 5.18 Preparation schemes for ion-exchange membranes based on high-density polyethylene (HDPE) film

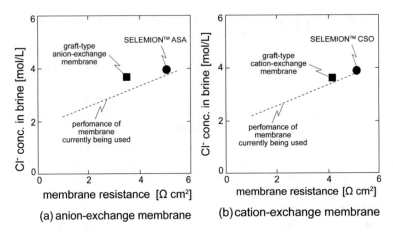

Fig. 5.19 Chloride ion concentration in brine versus membrane resistance in electrodialysis for graft-type ion-exchange membranes prepared by Chiba University versus SELEMION™ manufactured by AGC Engineering Co.

compared with currently used ion-exchange membranes for the electrodialysis of seawater. Date shown in this figure has helped improve confidence in this preparation scheme. In the near future, almost all edible salt in Japan will be manufactured by permeating seawater through the "super"-ion-exchange membranes prepared by radiation-induced graft polymerization and subsequent chemical modifications.

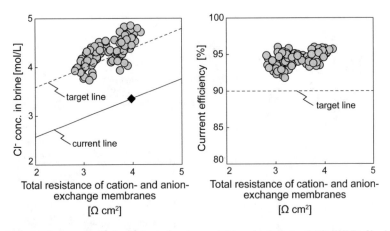

Fig. 5.20 Performance of ultrahigh molecular weight polyethylene (UHMWPE)-film-based ion-exchange membranes developed by The Salt Industry Center of Japan

References

1. T. Someya, Y. Okamura, G. Wada, Y. Shimoda, D. Umeno, K. Saito, N. Shinohara, N. Kubota, Comparison of resolution of proteins in elution chromatography between cation-exchange polymer brush immobilized particle-and commercially available cation-exchange-bead-packed beds. J. Ion Exch **24**, 1–7 (2013)

2. T. Sasaki, S. Uchiyama, K. Fujiwara, T. Sugo, D. Umeno, K. Saito, Similarity of rare earth extraction by acidic extractant bis(2-ethylhexyl) phosphate (HDEHP) supported on a dodecylamino-group-containing graft chain and by HDEHP dissolved in dodecane. Kagaku KogakuRonbunsyu. **40**, 404–409 (2014)

3. T. Sasaki, S. Uchiyama, K. Fujiwara, T. Sugo, D. Umeno, K. Saito, Nd/Dy resolution by SPE-based elution chromatography with bis(2-ethylhexyl) phosphate (HDEHP)-impregnated fiber-packed bed. Kagaku Kogaku Ronbunsyu **41**, 220–227 (2015)

4. N. Kubota, Y. Konno, S. Miura, K. Saito, K. Sugita, K. Watanabe, T. Sugo, Comparison of two convection-aided protein adsorption methods using porous membranes and perfusion beads. Biotechnol. Prog. **12**, 869–872 (1996)

5. K. Yamashiro, K. Miyoshi, R. Ishihara, D. Umeno, K. Saito, T. Sugo, S. Yamada, H. Fukunaga, M. Nagai, High-throughput solid-phase extraction of metal ions using an iminodiacetate chelating porous disk prepared by graft polymerization. J. Chromatogr. A **1176**, 37–42 (2007)

6. K. Ikeda, D. Umeno, K. Saito, F. Koide, E. Miyata, T. Sugo, Removal of boron using nylon-based chelating fibers. Ind. Eng. Chem. Res. **50**, 5727–5732 (2011)

7. K. Ikeda, D. Umeno, K. Saito, T. Kikuchi, K. Ando, T. Sugo, Selection of solvents suitable for immobilization of *N*-methylglucamine on poly(glycidyl methacrylate) grafted onto nylon fiber. J. Ion Exch **22**, 81–86 (2011)

8. Y. Matuzaki, T. Itabashi, S. Noma-Kawai, D. Umeno, K. Saito, Acrylic acid-grafted fiber enables high-capacity binding under high ionic strength of protein solution. Submitted to Radioisotopes (2017)

9. R. Kawamura, S. Goto, Y. Matsuura, S. Kawai-Noma, D. Umeno, K. Saito, K. Fujiwara, T. Sugo, Y. Yajima, A. Kinoshita, A. Kudo, J. Hioki, H. Wakabayashi, Adsorption of catechin in green-tea extracts onto NVP-grafted fiber and its elution with NaOH. Kagaku Kogaku Ronbunshu, in press (2017)

10. Y. Matsuura, R. Kawamura, S. Kawai-Noma, D. Umeno, K. Saito, K. Fujiwara, T. Sugo, Y. Yajima, J. Hioki, and H. Wakabayashi. Comparison of removal performance for catechin from green-tea extracts between NVP- and NVAA-grafted fibers. Submitted to Radioisotopes (2017)

11. M. Katagiri, M. Kono, S. Goto, K. Fujiwara, T. Sugo, S. Kawai-Noma, D. Umeno, K. Saito, Impregnation of sodium titanate onto DMAPAA-grafted fiber under mild reaction conditions and its strontium removal performance from seawater. Bull. Soc. Sea Water Sci., Jpn. **69**, 270–276 (2015)

12. S. Naruke, S. Goto, M. Katagiri, K. Fujiwara, T. Sugo, S. Kawai-Noma, D. Umeno, K. Saito, Determination of the composition and strontium-binding ratio of sodium titanate impregnated onto DMAPAA-grafted fiber. Bull. Soc. Sea Water Sci. Jpn. **70**, 364–368 (2016)

13. Y. Asari, T. Miyazawa, K. Miyoshi, D. Umeno, K. Saito, T. Nagatani, N. Yoshikawa, Development of novel ion-exchange membranes for electrodialysis of seawater prepared by electron-beam-induced graft polymerization (III) Co-graft polymerization of glycidyl methacrylate and divinylbenzene onto high-density polyethylene film. Bull. Soc. Sea Water Sci. Jpn. **63**, 387–394 (2009)

14. N. Yoshikawa, Report of Research institute of salt and sea water science. The Salt Industry Center of Japan, **10**, 25–28 (2008)

Chapter 6
Commercial Products by Radiation-Induced Graft Polymerization

Abstract Radiation-induced graft polymerization is a powerful tool for the following reasons: (1) From the macroscopic standpoint, the form of the adsorbent can be selected. For example, nonwoven fabrics and porous sheets may be adopted as trunk polymers instead of beads or granules. (2) From the microscopic standpoint, graft chains are relatively flexible, providing a novel space for ions and molecules. For example, proteins can be multilayered via multipoint binding, and inorganic precipitates can be immobilized through entanglement and penetration. (3) From an industrial standpoint, the pre-irradiation method is advantageous in that the processes, i.e., irradiation and grafting, are separable. An electron-beam-irradiated wound film and bobbins of gamma-ray-irradiated fibers can be used as trunk polymers in continuous and batch modes, respectively. Many polymeric adsorbents of various forms and components can be produced by radiation-induced graft polymerization.

Keywords Pre-irradiation grafting · Electron-beam-irradiated wound film Bobbin of gamma-ray-irradiated fiber

Companies that have manufactured the following six commercial products made from materials resulting from radiation-induced graft polymerization are listed in Table 6.1: (1) a diaphragm of a button-shaped silver oxide battery (Yuasa Corporation), (2) a filter for removing trace amounts of metal ions from ultrapure water (Asahi Kasei Chemicals Co.), (3) an anion-exchange porous hollow-fiber membrane for protein removal (Asahi Kasei Medicals Co.), (4) a nonwoven fabric adsorber of basic gases (ECE Co.), (5) iodine-impregnated nonwoven fabrics with antibacterial activity (KJK Co. and ECE Co.), and (6) an adsorptive fiber for removal of radioactive cesium ions from contaminated water (KJK Co.). These products are made in Japan from polymeric materials developed by many researchers and engineers engaged in the development of radiation-induced graft polymerization.

Table 6.1 Commercialized products of radiation-induced graft polymerization by private companies

	Trunk polymer	Vinyl monomer
Yuasa Corp.	PE film	AAc
Asahi Kasei Chemicals	PE hollow-fiber membrane	GMA
	PE particle	GMA
Asahi Kasei Medical	PE	GMA
ECE	PE/PP nonwoven fabric	CMS
KJK	nylon fiber	GMA, SSS, VBTAC

AAc acrylic acid; *GMA* glycidyl methacrylate; *CMS* chloromethyl styrene
SSS sodium styrene sulfate; *VBTAC* vinylbenzyl trimethylammonium chloride

6.1 Acrylic Acid-Grafted Diaphragm for Battery

Since the 1970s, the diaphragm of a silver oxide (Ag_2O) battery with a button shape has been produced by Yuasa Corporation. The production process was achievable under the intensive guidance of Takanobu Sugo, one of the authors of this book. This diaphragm, the first product manufactured by radiation-induced graft polymerization, is commercially available around the world. The product is a polyethylene-based wound film with grafted polyacrylic acid (AAc) chains, the width and diameter of which are 40 and 20 cm, respectively.

The structure of the button-shaped silver oxide battery is illustrated in Fig. 6.1. Conventionally, semipermeable diaphragms were used to separate two polar electrolyte solutions. However, in this structure, trace amounts of potassium hydroxide (KOH) transfer through the diaphragm from the electrolyte solution (40% KOH aqueous solution) on the Zn electrode side to that on the Ag_2O electrode side. The hydroxide ions react with the Ag_2O electrode to form $Ag(OH)_2^-$ ions. The resultant $Ag(OH)_2^-$ ions diffuse backward into the electrolyte solution of the Zn electrode through the diaphragm driven by its concentration gradient. The following reaction at the Zn electrode degrades the electrical performance and shortens the battery's life.

Fig. 6.1 Long-lasting button-shaped battery with an acrylic acid-grafted film as a diaphragm

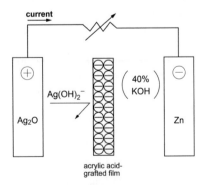

$$Zn + Ag(OH)_2^- \rightarrow Ag + ZnO + H_2O + e^-$$

Takanobu Sugo proposed that a cation-exchange membrane could very effectively replace the semipermeable diaphragm because the membrane blocks the permeation of hydroxide ions. To this end, AAc was graft-polymerized uniformly across the thickness of a high-density polyethylene (HDPE) film of 30 μm thickness. A small extent of swelling of the membrane is permissible in the directions of its thickness and surface because the film is incorporated into a button-shaped battery that is approximately 1 cm in diameter. The use of the AAc-grafted film as a thin cation-exchange membrane has greatly prolonged the battery's life.

On the basis of the laboratory data acquired from graft polymerization of AAc onto an electron-beam-irradiated HDPE film with an area of 5–10 cm^2, graft polymerization by pre-irradiation was scaled up. A wound HDPE film was irradiated with gamma rays at a commercial irradiation facility and stored in a freezer to prevent the decay of the radicals in the HDPE film. The wound irradiated film was transported to a production plant and installed at the starting point of a continuous graft polymerization unit. Then, the wound film was unwound and immersed into an AAc solution at a constant speed to ensure a prescribed degree of grafting.

As the graft polymerization of AAc progressed, the HDPE film swelled; therefore, special attention was given to minimizing the variation in the degree of grafting. The degree of grafting is controllable within an acceptable range to meet the requirements of electrical resistance and ion transport number for use as a diaphragm for a battery. This first commercial product from the technology associated with radiation-induced graft polymerization was presented with the Award from the Head Officer of Science and Technology of Japan in 1976.

6.2   Ion-Exchange Porous Hollow-Fiber Membranes

6.2.1   Metal Ion Removal for Ultrapure Water

To remove trace amounts of metal ions from water to produce ultrapure water, a porous sulfonic acid-containing cation-exchange hollow-fiber membrane was developed in 1997 by Hori et al. [1] of Asahi Kasei Chemicals Co. A porous polyethylene hollow-fiber membrane used for microfiltration of colloids and microorganisms was adopted as a trunk polymer for grafting. Styrene (St) was graft-polymerized onto a hollow fiber, and subsequently a sulfonic acid group was introduced into the benzene ring of the poly-St graft chain. During the permeation of water through the pores of the porous strongly acidic cation-exchange hollow-fiber membrane, trace amounts of metal ions, such as sodium and copper ions, were captured while simultaneously filtrating out particles suspended in the water, the removal of which is necessary to produce ultrapure water.

For application of the modified porous hollow-fiber membrane to ultrapure water production, the cost was high. Instead of a porous polyethylene hollow-fiber membrane, polyethylene particles were used as the trunk polymer. Sulfonic acid group-containing cation-exchange particles similarly prepared by radiation-induced graft polymerization were packed into a Baumkuchen-like cast and sintered. The resultant Baumkuchen-like cation exchanger captured metal ions during the flow of liquid through the interstices of the particles. Products with various inner and outer diameters of the Baumkuchen have been commercialized for the efficient removal of metal ions to produce ultrapure water by Asahi Kasei Chemicals Co. (Fig. 6.2). The technology for producing this Baumkuchen received the award from the Ion-Exchange Society of Japan in 2000.

6.2.2   Protein Removal from Biological Fluids

A porous anion-exchange hollow-fiber membrane for the removal of undesirable proteins from biological fluids was commercialized by Asahi Kasei Medical Co. in 2011 (Fig. 6.3). The anion-exchange moiety is the diethylamino group as a weakly basic anion-exchange group. This novel adsorbent was named QyuSpeed D [2]. A module charged with the hollow fibers has advantages over a bed charged with conventional anion-exchange beads in that the higher flow rate of a protein solution leads to a higher overall protein binding rate. This advantage arises because the

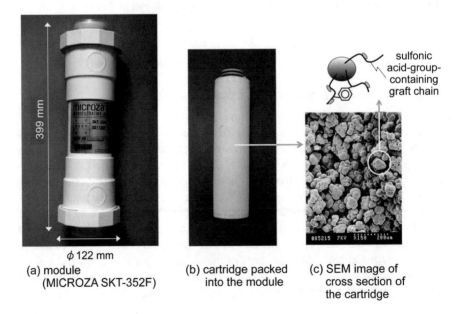

(a) module   (b) cartridge packed   (c) SEM image of
    (MICROZA SKT-352F)   into the module   cross section of
                                                                   the cartridge

Fig. 6.2 Module charged with sulfonic acid cation-exchange particles

mass-transfer resistance of proteins to the anion-exchange group during permeation through the pores rimmed by the graft chains is negligible. In addition, Kubota et al. [3] demonstrated the potential of a linear scale-up of the protein binding performance of porous hollow-fiber membranes by comparing between a single hollow fiber and a module consisting of eight hollow fibers. In contrast, a nonlinear scale-up of bead-packed beds is required to optimize the bed height and diameter for protein binding.

Approximately, twenty years has elapsed between discovery and commercialization of the product on the basis of fundamental studies by Tsuneda et al. [4] in 1990. Because this hollow-fiber membrane is applied to the purification of pharmaceuticals such as antibodies, concerns about safety, such as negligible detection of chemicals eluted from the membrane and protein adsorptivity, such as binding rate and capacity, must be addressed. In addition, durability against repeated use in adsorption and elution cycles and in pasteurization was evaluated. Various and severe requirements must be satisfied to apply the results of fundamental research to commercialization. As a result, however, many sizes of modules consisting of thousands of hollow-fiber membranes are now sold in the pharmaceutical market.

6.3 Nonwoven Fabric Adsorbers for Gases

In factories that manufacture semiconductors, the degree to which an ultraclean environment can be achieved in the workplace govern the yield of products. Gases coming from chemicals and workers should be removed with gas adsorbers to maximize the yield of semiconductors. Conventionally, air in an ultraclean room is circulated through an activated-carbon-packed bed to remove basic gases such as ammonia. However, activated carbon has a low binding capacity for ammonia, resulting in the need for a large volume of activated carbon and for its frequent regeneration. In addition, re-emission of basic gases may occur according to variations in ambient temperature because the principle of gas adsorption onto activated carbon is physical adsorption.

Fujiwara [5], one of the authors of this book, proposed the replacement of activated carbon with a sulfonic acid group-containing nonwoven fabric prepared

Fig. 6.3 Porous anion-exchange hollow-fiber membrane designed for removal of undesirable proteins in purification of pharmaceuticals

by radiation-induced graft polymerization. This novel gas adsorber based on the principle of chemical adsorption, i.e., neutralization, which was referred to as a chemical filter, is capable of removing several basic gases.

Radiation-induced graft polymerization is classified as either pre-irradiation or simultaneous irradiation based on the timing of the electron-beam or gamma-ray irradiation onto the trunk polymers. Industrially, pre-irradiation has an advantage over simultaneous irradiation in that the process of irradiation is separable from that of the graft polymerization and that the formation of homopolymers is minimized.

Irradiation by either electron beams or gamma rays is performed by private companies, and the trunk polymer containing radicals are stored in a freezer before transportation of the irradiated polymer to a factory for graft polymerization. To specialize in the technology of radiation-induced graft polymerization, in 2000, Ebara Corporation established the Ebara Clean Environment (ECE) Co., where a continuous unit for production of functional nonwoven fabrics by radiation-induced graft polymerization was constructed. The unit consists of equipment for electron irradiation and reactors for graft polymerization and subsequent introduction of functional groups (Fig. 6.4). This is the first continuous production line for the series of steps of pre-irradiation, graft polymerization, and functionalization.

N-Vinylpyrrolidone was graft-polymerized onto nonwoven fabrics, followed by reaction with iodine. The change in color of the nonwoven fabrics from white to brown is indicative of the complex formation of the pyrrolidone moiety with iodine. The complex is effective in killing bacteria because sublimed iodine destroys cell membranes. A mask made of the iodine-impregnated fibers was commercialized under the name of ISODINE mask (Fig. 6.5).

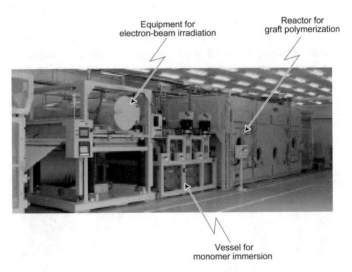

Fig. 6.4 Continuous production system for grafted materials consisting of equipment for electron-beam irradiation, a vessel for monomer immersion, and a reactor for graft polymerization

Fig. 6.5 Mask including iodine-impregnated nonwoven fabric

6.4 Adsorptive Fibers for Removal of Radioactive Substances

6.4.1 Mass Production of Cs-Adsorptive Fiber Using Batch Reactor

Within three months after the meltdown of three reactors of the TEPCO's Fukushima Daiichi Nuclear Power Plant, the Saito Laboratory at Chiba University prepared a procedure for manufacturing adsorptive fibers for the removal of cesium ions from water contaminated with radioactive substances. Using this procedure, KJK Co. and SUN-ESU Industry Co. succeeded in mass-producing bobbin-shaped adsorptive fibers [6]. Bobbins of approximately 90 kg in mass were produced per batch of a reactor. This mass production enabled a ready supply for the decontamination sites.

The hollow part of the bobbin pre-irradiated with gamma rays was inserted into a rod standing upright and perpendicular to the bottom of the reactor. The reactor contained 15 rods. One rod can hold six bobbins in series, as shown in Fig. 6.6. By permeating a monomer solution through the bobbins radially from inside to outside, a uniform degree of grafting over all bobbins was achieved. Then, functional groups were introduced followed by impregnation of cobalt ferrocyanides in the permeation mode. From the resultant bobbins, various forms of fibers, such as braids and wound filter, were manufactured.

Fig. 6.6 Reactor for
radiation-induced graft
polymerization capable of
processing 200 fiber-bobbins
per batch

Bobbin

The reasons for difficulties in the removal of radionuclides from liquids are as
follows: (1) Some radionuclides dissolve at extremely low concentrations in water;
therefore, inorganic substances capable of incorporating radionuclides into a crystal
lattice and between layers of inorganic substances are more effective than
ion-exchange and chelate-forming groups. (2) Adsorbents cannot distinguish
radioactive ions from nonradioactive ions; when radionuclides are added to originally dissolved nonradionuclides, high binding capacity becomes an essential
requisite for adsorbents. (3) Radionuclide-bound adsorbents must be stored as
radioactive waste for a long period; therefore, volume reduction of the adsorbent by
incineration is desirable. Insoluble cobalt-ferrocyanide-impregnated fibers exhibit a
high affinity for cesium ions in seawater, and volume reduction is possible by
incinerating both trunk polymer (nylon 6) and grafted polymer chain. This technology for mass-producing Cs-adsorptive fiber received the Resona SME excellent
new technology and new product Award in 2015.

6.4.2   Difficulty of Radioactive Strontium Removal

Removal of radioactive strontium ions from contaminated seawater stored in harbors
and contaminated water stored in tanks is more difficult than the removal of
radioactive cesium ions for the following reasons: (1) radioactive strontium released
following serious accidents at a nuclear power plant dissolves in seawater where
nonradioactive strontium originally dissolves at a concentration of approximately
8 mg/L; the concentration of cesium originally dissolved in seawater is 0.0003 mg/L.
Therefore, any adsorbent must collect not only radioactive strontium but also nonradioactive strontium. (2) Alkaline-earth metals similar to strontium dissolve in
seawater at much higher concentrations than strontium, such as magnesium and

calcium at 1400 and 400 mg/L, respectively. The molar concentrations of magnesium and calcium are 640- and 110-folds higher than that of strontium. For radioactive strontium, even nonradioactive strontium in addition to magnesium and calcium are regarded as competing ions in seawater.

Sodium titanate is a promising candidate with respect to affinity for strontium ions even in seawater. A commercially available adsorbent, SrTreat [7], manufactured by Fortum Co. in Finland, contains sodium titanate. We impregnated sodium titanate onto nylon-6 fibers by radiation-induced graft polymerization and subsequent chemical modifications. Now, we are planning to increase the amount of sodium titanate impregnated and to maximize the binding capacity for strontium even in seawater [8–10].

6.4.3 Proposal for Decontamination Using Adsorptive-Fiber Assembly

The TEPCO's Fukushima Daiichi Nuclear Power Plant (NPP) includes two types of water contaminated with radionuclides (Fig. 6.7): (1) contaminated water stored in tanks. The volume of stored water amounted to one million tons (1,000,000 m^3) by the end of March in 2017 and (2) contaminated seawater in the harbor in front of the seawater-intake area of reactors 1–4. The volume is calculated to be 160,000 m^3 because the length, width, and depth of the harbor are 400, 80, and 5 m, respectively. These two types of water are expected to undergo treatment for radionuclide

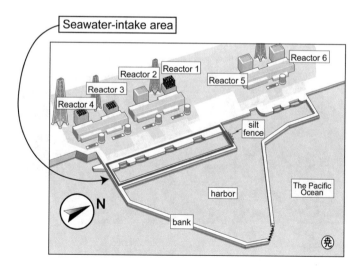

Fig. 6.7 Seawater-intake area in front of reactors 1–4 at TEPCO's Fukushima Daiichi Nuclear Power Plant

A braid of 80-m-long Cs adsorptive fiber

Fig. 6.8 Immersion of braids of Cs-adsorptive fibers into seawater-intake pit in front of Reactor 3 of TEPCO's Fukushima Daiichi Nuclear Power Plant

removal before they are released into the ocean with the permission of the fishermen's association.

Equipment designed for the removal of various radionuclides from contaminated water in storage tanks has been installed at the TEPCO's Fukushima Daiichi NPP. The equipment is referred to as the advanced liquid processing system (ALPS). The pretreatment unit in ALPS processes the contaminated water by coagulation and precipitation with iron ions and carbonates, respectively, generating a large volume of radioactive slurry that must be stored over a long period. The coagulation and precipitation operations have now been partly replaced by a filtration operation using wound filters consisting of cesium- and strontium-adsorptive fibers [11, 12]. These cesium- and strontium-adsorptive fibers are cobalt-ferrocyanide-

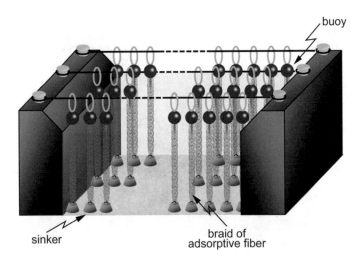

Fig. 6.9 Illustration of Cs and Sr removal from seawater-intake area in front of Reactors 1–4 using aligned braids of adsorptive fibers

| (a) Braids of Cs- and Sr-adsorptive fibers were aligned in curtain form | (b) The curtain was immersed in the seawater in the harbor of the TEPCO Fukushima Daiichi Nuclear Power Plant |

Fig. 6.10 Immersion of braids of Cs- and Sr-adsorptive fibers in curtain form into seawater contaminated with radionuclides

impregnated and iminodiacetate-containing chelating fibers, respectively, prepared by radiation-induced graft polymerization and subsequent chemical modifications. A newly modified ALPS is expected to reduce radioactive waste by 90% compared with the conventional ALPS because no radioactive slurry is generated. The nuclide-bound wound filters will be easily stored in high-integrity containers (HIC).

In June 2013, a braid of cobalt-ferrocyanide-impregnated fibers was immersed in the pit of the seawater-intake area of the TEPCO's Fukushima Daiichi NPP to demonstrate its cesium removal performance, as shown in Fig. 6.8. To remove nonradioactive strontium as well as radioactive strontium dissolved in a volume of seawater of 160,000 m^3, a tremendous mass of fibers as adsorbent is required. As illustrated in Fig. 6.9, an assembly of braids was proposed. An adsorbent in fiber

(a) Bird's eye view of TEPCO Fukushima Daiichi NPP

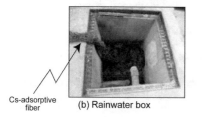
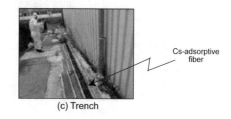

Cs-adsorptive fiber

Cs-adsorptive fiber

(b) Rainwater box   (c) Trench

Fig. 6.11 Cs-adsorptive fibers installed in various sites at TEPCO's Fukushima Daiichi NPP

form has an advantage over that in bead form in that the immersion and recovery of the adsorbent are easily managed and that the volumes of Cs- and Sr-bound fibers can be easily reduced by incineration because components except for Cs-bound cobalt ferrocyanide and Sr-bound sodium titanate are organic polymeric materials.

The fiber has the basic merit of rapid adsorption of a target radionuclide because of a high specific external surface area and a low diffusional mass-transfer resistance. In December 2014, a number of cobalt-ferrocyanide-impregnated braids were attached to a curtain wall and the wall was immersed in the harbor to confirm the removal of cesium and strontium ions from seawater (Fig. 6.10). In addition, the braids were installed at various locations on-site such as drainage areas, rainwater boxes, and trenches (Fig. 6.11) [13].

References

1. T. Hori, M. Hashino, A. Omori, T. Matsuda, K. Takasa, K. Watanabe, Synthesis of novel microfilters with ion-exchange capacity and its application to ultrapure water production systems. J. Membr. Sci. **132**, 203–211 (1997)
2. H. Shirataki, C. Sudoh, T. Eshima, U. Yokoyama, K. Okuyama, Evaluation of anion-exchange hollow-fiber membrane adsorber containing γ-ray grafted glycidyl methacrylate chains. J. Chromatogr. A **1218**, 2381–2388 (2011)
3. N. Kubota, Y. Konno, K. Saito, K. Sugita, K. Watanabe, T. Sugo, Module performance of anion-exchange porous hollow-fiber membranes for high-speed protein recovery. J. Chromatogr. A **782**, 159–165 (1997)
4. S. Tsuneda, K. Saito, S. Furusaki, T. Sugo, High-throughput processing of proteins using a porous and tentacle anion-exchange membrane. J. Chromatogr. A **689**, 211–218 (1995)
5. K. Fujiwara, *Ebara Engineering Review*, vol. 146 (1990), pp. 1–7
6. Y. Okamura, K. Fujiwara, N. Iijima, T. Syoda, K. Suzuki, T. Sugo, T. Shimidu, R. Itagaki, A. Takahashi, T. Ono, T. Kikuchi, T. Someya, R. Ishihara, T. Kojima, D. Umeno, K. Saito, Preparation of adsorptive fibers for removal of cesium from seawater. J. Ion Exchange **24**, 8–13 (2013)
7. E. Tusa, in *Efficiency of Fortum's CsTreat^{TM} and SrTreat^{TM} in cesium and strontium removal in Fukushima Daiichi NPP, Homepage of Fortum Co*
8. M. Kono, S. Umino, S. Goto, K. Fujiwara, T. Sugo, T. Kojima, S. Kawai-Noma, D. Umeno, K. Saito, Preparation of adsorptive fiber by a combination of peroxo complex of titanium anion and DMAPAA-grafted fiber for the removal of strontium from seawater. Bull. Soc. Sea Water Sci. Jpn. **69**, 90–97 (2015)
9. S. Goto, M.J. Katagiri, S. Naruke, K. Fujiwara, T. Sugo, T. Kojima, S. Kawai-Noma, D. Umeno, K. Saito, in *Linear relationship between impregnation percentage of sodium titanate of adsorptive fiber and adsorption capacity for strontium in artificial seawater*. Submitted to *Bunseki Kagaku*
10. S. Goto, M. Katagiri, S. Naruke, K. Fujiwara, T. Sugo, S. Kawai-Noma, D. Umeno, K. Saito, in *Improvement in impregnation percentage of sodium titanate of adsorptive fiber through repetitive immobilization of peroxotitanium complex anions to anion-exchange fiber*. Submitted to *Bunseki Kagaku*
11. R. Ishihara, K. Fujiwara, T. Harayama, Y. Okamura, S. Uchiyama, M. Sugiyama, T. Someya, W. Amakai, S. Umino, T. Ono, A. Nide, Y. Hirayama, T. Baba, T. Kojima, D. Umeno, K. Saito, S. Asai, T. Sugo, Removal of cesium using cobalt-ferrocyanide-impregnated polymer-chain-grafted fibers. J. Nucl. Sci. Technol. **48**, 1281–1284 (2011)

12. S. Goto, S. Umino, W. Amakai, K. Fujiwara, T. Sugo, T. Kojima, S. Kawai-Noma, D. Umeno, K. Saito, Impregnation structure of cobalt ferrocyanide microparticles by the polymer chain grafted onto nylon fiber. J. Nucl. Sci. Technol. **53**, 1251–1255 (2016)
13. TEPCO, in *Evaluation of performance of Cs and Sr adsorptive fibers installed in the harbor of TEPCO Fukushima Daiichi NPP* (2016, 22 Dec)

Epilogue

Examples of Countermeasures Against Difficulties of Separation

To remove trace amounts of cesium and strontium ions from contaminated water, insoluble cobalt ferrocyanide and sodium titanate were immobilized onto a commercially available nylon-6 fiber. At first, strongly basic anion-exchange-group-containing graft chains were added to the nylon fibers. A "graft phase" consisting of these graft chains provides space for the formation of precipitates of insoluble cobalt ferrocyanide and sodium titanate. The resultant inorganic-compound-impregnated fiber was capable of capturing cesium and strontium ions in seawater based on an ion-exchange interaction.

To remove urea to make ultrapure water, urease was immobilized onto the nylon fiber to decompose urea into ammonium ions and carbon dioxide. At first, a weakly basic anion-exchange-group-containing graft chain was bonded to the nylon fiber. A "graft phase" consisting of ionizable graft chains provided an enzyme-multilayering space. Subsequently, urease-multilayered graft chains were enzymatically crosslinked with transglutaminase to prevent the elution of urease from the graft chains caused by pH changes during the progress of hydrolysis of urea. The resultant urease-immobilized fiber was capable of hydrolyzing urea at a high rate.

To separate and concentrate neodymium (Nd) and dysprosium (Dy) ions in acidic media, an acidic extractant, HDEHP, was impregnated into the nylon fiber. Dodecylamine was reacted with the glycidyl methacrylate graft chain bonded to the nylon fiber. A "graft phase" consisting of hydrophobic graft chains provided a space for dissolving extractant that contains both hydrophobic and ionizable parts. The extractant-impregnated fiber was capable of resolving Nd and Dy ions by elution chromatography.

© Springer Nature Singapore Pte Ltd. 2018
K. Saito et al., *Innovative Polymeric Adsorbents*,
https://doi.org/10.1007/978-981-10-8563-5

Changes in Role of Graft Chains from Central Player to Supporting Player

Key functional components in polymeric adsorbents capable of removing radionuclides and urea or of purifying rare-earth metals are cobalt ferrocyanide, urease, and HDEHP, respectively. Graft chains support these components. To data, according to applications, we have graft-polymerized an epoxy-group-containing vinyl monomer onto a porous hollow-fiber membrane, a porous sheet, and a non-porous film, followed by the introduction of functional groups such as ion-exchange and chelate-forming groups and hydrophobic and affinity ligands. The added functional groups mainly govern the performance in separations. However, they are not sufficient for overcoming the difficulties in the separation of radionuclides in decontamination, procedures in the production of ultrapure water, and in the purification of rare-earth metals.

Some inorganic compounds can specifically capture cesium ions at extremely low concentrations. Enzymes decompose substrates at extremely low concentrations. Extractants separate elements with very similar properties. These components must be immobilized onto solid supports at high densities without degrading their functionality; therefore, a flexible space is desirable for the immobilization of components. Ionizable-group-containing graft chainscan extend themselves because of mutual electrostatic repulsion to form spaces that are accessible to components. Inorganic compounds and enzymes have been immobilized on ionizable graft chains. In addition, extractants have been impregnated or immobilized onto hydrophobic graft chains that then behave like organic solvents.

Keypoints in Revolution of Polymeric Adsorbents by Radiation-Induced Graft Polymerization

Accessible forms of polymeric adsorbents include beads, granules, fibers, hollow-fiber membranes, sheets, nonwoven fabrics, and films, regardless of whether the inner structure is porous or nonporous. Existing polymers have readily been modified by the radiation-induced graft polymerization of reactive vinyl monomers onto them and subsequent chemical reactions. An appropriate selection of the form of a polymeric adsorbent contributes to an increase in the overall adsorption rate of ions and molecules. Hydrophilic or ionizable graft chainscan extend themselves, providing a space for incorporating inorganic compounds and enzymes. In contrast, hydrophobic graft chains do not stretch, providing a solvent-like medium for dissolving extractants. High capacity and high durability of immobilized components were achievable.

Inorganic compounds, enzymes, and extractants as well as ion-exchange and chelate-forming groups and hydrophobic and affinity ligands have some advantages over conventional adsorbents because they are immobilized at high densities onto various trunk polymers that possess large external surface areas and low mass-transfer resistance. Thus, radiation-induced graft polymerization has revolutionized polymeric adsorbents. In addition, on the basis of procedures prescribed by university laboratories, e.g., the Saito laboratory of Chiba University, private companies, e.g., KJK Co., have completed the scale-up of graft polymerization and the introduction of functional moieties. Pre-irradiation techniques that can store the radicals produced by electron beam or gamma ray irradiation are effective for the mass production of polymeric adsorbents because the irradiation process and grafting process are separable.

Printed in the United States
By Bookmasters